FABRICATION of SILICON MICROPROBES for OPTICAL NEAR-FIELD APPLICATIONS

Phan Ngoc Minh
Ono Takahito
Esashi Masayoshi

CRC PRESS

Boca Raton London New York Washington, D.C.

Library of Congress Cataloging-in-Publication Data

Minh, Phan Ngoc.
    Fabrication of silicon microprobes for optical near-field applications / Phan Ngoc Minh,
Ono Takahito, Esashi Masayoshi.
       p. cm.
Includes bibliographical references and index.
ISBN 0-8493-1154-3 (alk. paper)
    1. Micromachining. 2. Silicon. 3. Near-field microscopy. 4. Microprobe analysis. I.
Takahito, Ono. II. Masayoshi, Esashi. III. Title.

TJ1191.5.M5526 2002
621.36—dc21   2001043935

Visit the CRC Press Web site at www.crcpress.com

Preface

Near-field optics (NFO) currently is one of the most exciting areas of nano-optics and its related technologies using the creation of a bright, nanoscale light source. The most critical part of near-field optics is the microprobe. In this study a nanosized light source at a subwavelength aperture was used.

Currently a tapered metallized optical fiber probe with a sharp tip and a subwavelength optical aperture are most widely used. The performance of the fiber-based probes has been improving and such probes are commercially available. Their optical and mechanical properties, however, are poor and limit many applications with near-field light. The fabrication process of the fiber-based probes is not suited for mass production. Fiber-based probes must be produced and inspected individually, and are, therefore, expensive and less reproducible. Optical throughput of fiber-based probes is low, especially for probes with sub-100 nm aperture. It is difficult to utilize optical fiber-based probes in several applications where an array of nanoscale light sources is required. A technological solution for fabrication of high performance microprobes for optical near-field applications remains a problem.

In this work, several technological approaches using silicant (Si) micromachining technology for fabrication of microprobes for optical near-field and other applications are presented. A simple and effective technique was found for fabrication of Si cantilevered probes having an aperture as small as 20 nm in diameter, and its array in a batch process. Optical performances of the fabricated probes such as optical throughput, polarization behaviors, near-field spatial distribution were also investigated. Fabricated probes were found to overcome almost all drawbacks of conventional optical fiber-based probes in production throughput, optical throughput, reproducibility, simplicity of instrumentation, and mechanical performance.

Several novel structures for locally enhancing the near-field light, such as aperture with a nanometallic wire (coaxial structure), a single carbon nanotube (CNT), or a small Ag particle as a scatterer located at the center of the aperture, were proposed and fabricated in a batch process. A simulation of the finite difference time domain (FDTD) was carried out for the fabricated apertured tip. Optical near-field and corresponding atomic force images of several surfaces with a resolution of approximately 15 nm were demonstrated with the fabricated probes. First experimental results of optical near-field recording and reading of bits on a phase change medium (GeSbTe) were also performed using the fabricated aperture array.

Using the newly developed technique, other applications such as nanoscale thermal profiler, nanoscale heater for data storage, hybrid structure of optical fiber and apertured cantilever have been developed. The technique is also applicable in fields such as emission devices, nano Raman and nano Auger.

There are a number of excellent comprehensive books on near-field optics and Si microfabrication. This book is meant to introduce and suggest several technological approaches using Si micromachining technology to make a tool for working with near-field nanooptics, nanoscale processing and analyzing. The contents of this book are based on experimental research done at Tohoku University. It is the authors' hope that this book may be a useful reference to those working in the field of nanotechnology.

Phan Ngoc Minh
Ono Takahito
Esashi Masayoshi
Tohoku University
Sendai, Japan

Acknowledgments

The authors thank Dr. S. Tanaka of Tohoku University for his collaboration on simulation work. The following are acknowledged for their invaluable help: Prof. K. Hane of Venture Business Laboratory, Tohoku University; Prof. M. Ohtsu and Dr. T. Yatsui of Tokyo Institute of Technology; Prof. S. Kawata of Osaka University; Dr. J. Tominaga of the National Institute for Advanced Interdisciplinary Research, Tsukuba; Prof. K. Goto of Tokai University; Prof. E. Oesterschulze of Kassel University; Prof. P. H. Khoi and Dr. L. T. T. Tuyen of the Institute of Materials Science (Vietnam); and Dr. Y. Haga, Dr. T. Abe, Dr. K. Nakamura, H. Miyashita, H. Watanabe, and S. Shiba of Tohoku University.

Special thanks for their support: Esashi Noriko, Ono Mika, Nguygen Thi Ngoc Diep, and P. N. Minh's daughter Phan Ngoc Nhat Mai.

About the Authors

Phan Ngoc Minh holds a B.S. in physics from Hanoi National University (1991), a Ph.D. in physics from the Institute of Physics, Vietnam National Center for Natural Science and Technology (1996), and a Ph.D. in mechatronics and precision engineering from Tohoku University (2001). He was a principle investigator in the Institute of Materials Science, Vietnam National Center for Natural Science and Technology. He currently is a research fellow at the Venture Business Laboratory, Tohoku University. His current research interests include near-field optics, field emission devices, micro- and nano-fabrication for nanoprocessing, analyzing, and data storage.

Ono Takahito received a B.S. in 1990 in physics from Hirosaki University, an M.S. in 1992 in physics from Tohoku University, and a D.E. in 1996 in mechatronics and precision engineering, Tohoku University. Since 1996 he has been a research associate, and currently is an associate professor in the Department of Mechatronics and Precision Engineering, Tohoku University. His research interests are nanofabrication and nanomechatronics.

Esashi Masayoshi received a B.S. in electronic engineering in 1971 and a D.E. in 1976 at Tohoku University. He studied ion-sensitive field effect transistors and their packaging. From 1976 to 1981, he served as a research associate at the Department of Electronic Engineering, Tohoku University, and was an associate professor from 1981 to 1990. During this period, he worked on microsensors and integrated microsystems. He was a professor at the Department of Mechatronics and Precision Engineering, Tohoku University, from 1990 to 1998. He worked on micromachining and nanomachining. Since 1998, he has been a professor at the New Industry Creation Hatchery Center at Tohoku University. He is currently the director of the Venture Business Laboratory and an associate director of the Semiconductor Research Institute at Tohoku University. His current research topic is microtechnology for saving energy and natural resources.

Contents

chapter one

Introduction

1.1 *Introduction*

With the invention of scanning tunneling microscopy (STM) by Binnig and Rohrer in 1982,[1] people could touch atoms on a conductive surface for the first time. Since then a family of novel microscopes based on scanning probe microscopy (SPM) has been developed for local characterization, modification and investigation of the topographic, electrical, magnetic, mechanical, thermal and optical properties of surfaces at nano and atomic scales. Details of the development of SPM techniques have appeared in many publications since 1982. SPM is a marvelous tool for touching the "nanoworld" — certainly one of the most interesting areas of study in the 21st century.

In optical microscopy it is well known that the spatial resolution or spot size of a focused light beam is limited by the wavelength λ of the incident light and the numerical aperture (NA) of the optical system due to the diffraction effect. The diffraction effect limits every application that uses light, such as microscopy, spectroscopy, photolithography and optical data storage. For a long time, conventional optics seemed to be limited by the wavelength of the light due to the diffraction effect.

Fortunately, the diffraction effect can be overcome with the new concept of near-field optics (NFO). More than 70 years ago, Synge[2] proved that the diffraction effect can be overcome in optical microscopy by scanning a sub-wavelength-sized light source generated at a small aperture on a metallic screen in the sufficient proximity of a sample and detection of the diffracted light. With such a system, an optical resolution of approximately the aperture diameter can be achieved. In 1972 Ash and Nicholls[3] reported the first experimental study on near-field microscopy in the microwave region that supported Synge's idea.

Soon after the birth of STM and the development of the scanning technique, the first sub-wavelength resolution optical microscopy was independently found by Pohl et al. at IBM, Switzerland[4] and Lewis et al. at Cornell University[5] by utilizing a nanoscale light source generated at a sub-wavelength-size aperture at the apex of a metal-coated, tapered glass rod or tapered glass capillary micropipette. Since then, to create and utilize a bright light source with sub-wavelength-sized, namely *near-field* or *evanescent-field*, has opened up a new frontier in optics — near-field optics, nanooptics and related technologies. Presently near-field optics is one of the most exciting matters in optics. Novel optical microscopy using near-field light and related applications is referred to by several names: near-field scanning optical microscopy (NSOM), scanning near-field optical microscopy (SNOM), or photon scanning tunneling microscopy (PSTM).

Tapered optical fiber with a small optical aperture currently is used as the optical near-field probe. Although improvements have been achieved in the fabrication of fiber probes and they are commercially available, problems still remain. It is difficult to fabricate the fiber probe in a batch production with high reproducibility, both in the size and the shape of the aperture and tip. The optical fiber tip is fragile and optical transmission efficiency of the light through the aperture is low and limits many optical near-field applications. Moreover, it is difficult to utilize optical fiber probes in several applications that require an array of near-field light sources.

Silicon (Si) micromachining technology, however, is very advantageous. It has the capabilities of batch fabrication, precise controlling, integration, multifunctionality and miniaturization. Si micromachining has been the most appropriate technology for fabrication of microsensors, actuators, and microsystems, as well as microscanning probes.[6,7] Si microfabrication should be suitable for fabrication of high performance NSOM probes.

Several excellent books by outstanding experts in the field of Si micromachining technology, as well as near-field optics, including theory, probe fabrication, applications are available. This book does not cover all of these topics, but introduces several technological approaches to fabrication of an ultra-small aperture on a Si cantilever using Si micromachining technology for a hybrid atomic force microscopy (AFM)/NSOM and capacitive–AFM/NSOM probes. Several types of novel apertures for locally enhancing near-field light are also proposed here. An investigation of the optical performance of fabricated structures, both experimental and simulation, is presented. Several applications for optical near-field imaging, lithography, and optical near-field recording using the fabricated probes and the fabricated aperture array are also demonstrated. Instrumentation using the well-developed AFM technique for near-field optical applications is presented. This fabrication technique allows for the possibility of the development of several novel probes, such as a nanoheater integrated probe and its array for thermal imaging and recording, and an NSOM probe with an integrated waveguide or hybrid optical fiber and apertured cantilever. The technological solutions presented in this book can be effective, not only for optical near-field applications, but for other fields of nanotechnology as well.

1.2   *Structure of the book*

This book consists of 11 chapters with 130 illustrations and over 200 references. The structure of the book is as follows:

- Chapter 1 introduces the motivation and structure of the book.
- Chapter 2 presents a short introduction of near-field optics, instrumentation of optical near-field imaging, optical fiber-based NSOM probes, problems remaining with such fiber probes and our motivation with Si micromachined probes.
- Chapter 3 briefly introduces some fundamental techniques of Si microfabrication.
- Chapter 4 describes technological solutions for the fabrication of ultra-small aperture at the apex of a SiO_2 tip on a Si cantilever, both for hybrid AFM/NSOM, capacitive-AFM/NSOM probes, as well as aperture array for optical near-field recording head.
- Chapter 5 presents the optical performance, including the optical throughput, spatial distribution of near-field light at the aperture, polarization behaviors, as well as mechanical properties (static and dynamic behaviors) of the fabricated probes.
- Chapter 6 suggests several novel structures, such as coaxial tips with a nanometal wire-like, Ag particles or single carbon nanotube (CNT) at the center of the aperture. These novel probes are considered as candidates for improving the optical throughput by locally enhancing the intensity of the near-field light at the aperture. A development for the fabrication of a nanometallic contact at the aperture for thermal profiler and thermal recording probes, as well as hybrid structure of optical fiber or integrated waveguide with the apertured cantilever, are also presented in this chapter. A simple technique for fabrication of micro lenses at the core of an optical fiber-bundle for optical near-field and other applications is described. A primary result of the fabrication of electron field emission devices for multi electron beam lithography is also presented.
- Chapter 7 presents simulated results using a finite difference-time domain (FDTD) simulation. Distribution of near-field light at the aperture and optical throughput of the fabricated aperture were estimated with the FDTD simulation.
- Chapters 8, 9, and 10 demonstrate optical near-field imaging, lithography and optical near-field recording and reading for data storage using the fabricated structures and the developed measurement setups.
- Chapter 11 refers several further application possibilities, future aspects, and conclusions of the work.

References

1. Binnig, G. and Rohrer, H., Scanning tunneling microscopy, *Helv. Phys. Acta,* 55, 726, 1982.
2. Synge, E.H., A suggested model for extending microscopic resolution into the ultra-microscopic region, *Phil. Mag.,* 6, 356, 1928.
3. Ash, E.A. and Nichols, G., Super-resolution aperture scanning microscope, *Nature,* 237, 510, 1972.
4. Pohl, D.W., Denk, W., and Lanz, M., Optical stethoscopy: image recording with resolution $\lambda/20$, *Appl. Phys. Lett.,* 44, 651, 1984.
5. Harootunian, A. et al., Super resolution fluorescence near-field scanning optical microscopy, *Appl. Phys. Lett.,* 49, 674, 1986.
6. Madou, M., *Fundamentals of Microfabrication,* CRC Press, Boca Raton, FL, 1997.
7. Kovacs, G.T.A., *Micromachined Transducers: Sourcebook,* McGraw-Hill, New York, 1998.

part 1

Backgrounds

chapter two

Introduction to near-field optics

2.1 Far-field light and diffraction effect

Visible light is electromagnetic radiation with a wavelength of between 390 nm (violet) and 780 nm (red). Important parameters of electromagnetic radiation are wavelength (λ), frequency (ν) and energy (ε). Light is an electromagnetic wave but can also be viewed as a stream of discrete wave packets of energy called *photons*. The relationship between the wavelength, frequency and photon energy are expressed as:

$$\nu = c/\lambda \tag{2.1}$$

$$\varepsilon = h\nu = hc/\lambda \tag{2.2}$$

Where, $c \approx 3 \times 10^8$ m/s is the speed of light in a free space; $h = 6.67 \times 10^{-34}$ J.s is Planck constant.

If the wavelength of light is small in comparison with the size of the optical apparatus, corresponding to the limiting case $\lambda \rightarrow 0$, one can use as a first approximation, the techniques of "ray optics" or "geometrical optics".[1] However, if the dimensions of the system are small compared to the wavelength, a more precise treatment with "physical optics" should be used, i.e. the solution for the electromagnetic wave must be treated.

The wave nature of light in free space is well expressed by Maxwell's equations:[2]

$$\nabla \times E = -\partial B/\partial t \tag{2.3}$$

$$\nabla \times B = \varepsilon_o \mu_o \partial E/\partial t \tag{2.4}$$

$$\nabla \cdot B = \nabla \cdot E = 0 \tag{2.5}$$

Where E and B are the electric field and magnetic field vectors of the electromagnetic wave, respectively; ε_o, μ_o are the permittivity and permeability of the free space, respectively.

Many centuries ago, light was used for optical microscopy. With an optical microscope a person could clearly observe tiny objects that the human eye could not. Optical microscopy has been become a routine and indispensable tool in all areas of research, such as biology, medicine and materials. No other tool can replace the optical microscope. A simplified illustration of an ordinary optical microscope is shown in Figure 2.1. A light source (normally a white light) is shaped through an optics system and illuminates a sample. This is reflected from or transmitted through the sample and collected by an objective lens that enters a charge-couple device (CCD) camera or human eyes through an eyepiece. Optical microscopes allow one to clearly observe a tiny circumstance or object with a resolution of approximately one micrometer. The resolution or resolving power of the optical microscope refers to the distance between the two nearest circumstances on the object or the minimum size of the circumstance that can be clearly distinguished by the microscope. As shown in Figure 2.1 and will be discussed more detail later, when the distance between two circumstances A and B on the sample becomes smaller than a critical value (Rayleigh criterion), the distinction between circumstances of A and B by the CCD camera or the human eye is ineffective. To improve the resolution, the objective lens must shape the light beam to a smaller spot.

With the invention of laser (light amplification by stimulated emission of radiation) in 1960, new applications using light have been explored, such as in communications and optoelectronics. An example of these applications is the optical data storage family of compact-disks read only memory (CD-ROM), compact-disks rewritable (CDR), digital video disks (DVD) and

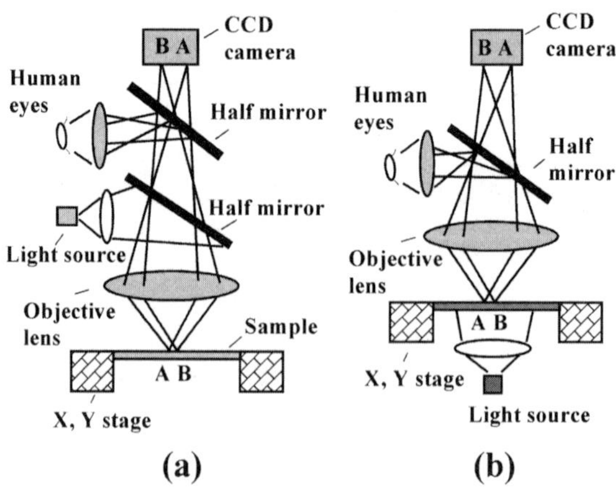

Figure 2.1 Simplified illustration of an ordinary optical microscope (a) reflection (b) transmission type.

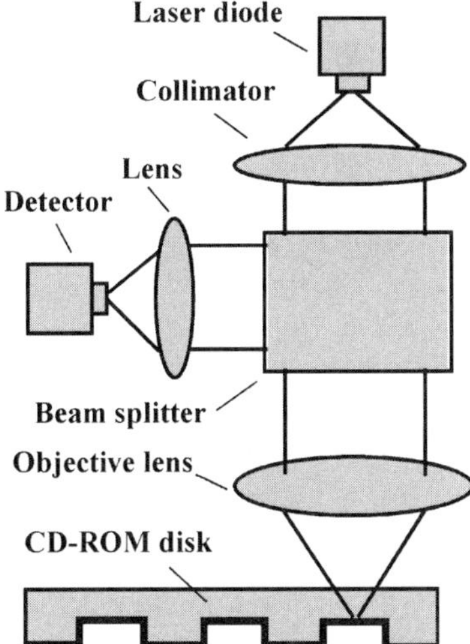

Figure 2.2 Simplified illustration of the optical compact disk read only memory (CD-ROM).

digital video recordables (DVR). A simplified illustration of the CD-ROM is shown in Figure 2.2. A focused laser beam passing through an objective lens is used to search and read-back prestamped bits on the CD disk. Since the bit's size and the distance between the two nearest bits must not be smaller than the beam spot size, bit's density of the optical disk is limited (it currently is 650 MB/disk). To increase the bit's density or reduce bit size, it is necessary to reduce the size of the focused light spot to read the data.

Unfortunately due to the fundamental nature of the electromagnetic wave, the spot size (d) of a focused light beam by an objective lens is limited by the wavelength of the incident light due to the diffraction effect as expressed in the equations below:

$$d \geq 1.22\ \lambda/\mathrm{NA} \tag{2.6}$$

$$\mathrm{NA} = n\sin(\theta) \tag{2.7}$$

where, λ is wavelength of the incident light, NA is the numerical aperture of the objective lens, n is the refractive index of the medium in the object space (normally, the medium between the lens and specimen is air, n = 1), θ is the semiaperture that is the half-angle of the maximum cone of light picked up by the objective lens (see Figure 2.3). The intensity of the light at the focused position has a spectrum as shown in the lower part of Figure 2.3 (Fourier transform spectrum). The spot size (d) of the focused laser beam approximates wavelength λ.

The diffraction limitation was pointed out by Abbe and Rayleigh more than 100 years ago and named the Abbe or Rayleigh criterion. The Rayleigh

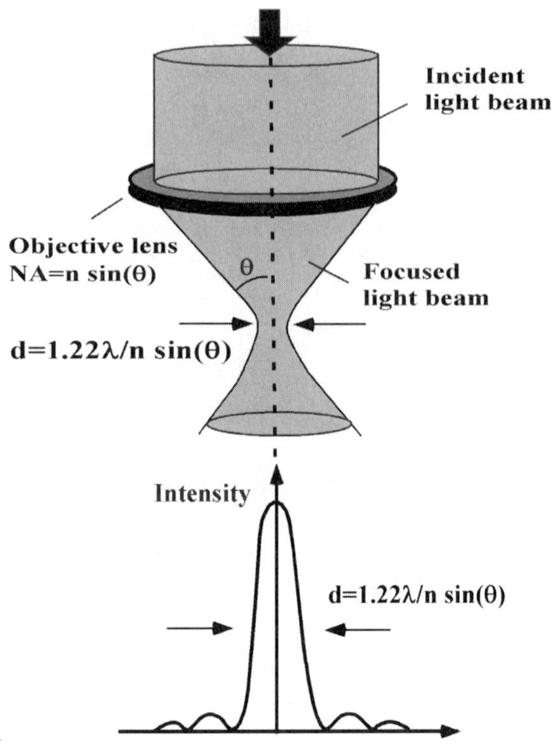

Figure 2.3 Spot size of a light beam through an optic system, d = 1.22 λ/nsinθ.

criterion for the resolution of two fringes with equal flux density requires that the principal maximum of one coincide with the first minimum of the other. As shown in Figure 2.4, when the distance of two circumstances (A and B of Figure 2.1) is larger than the Rayleigh criterion, the images can be clearly resolved. When the distance of two circumstances decreases below the Rayleigh criterion, the images overlap and are not resolved. This means that resolution of the optical microscope (d) is approximately half of wavelength λ and expressed as Equation 2.8:

$$d \geq 0.61 \, \lambda/NA \tag{2.8}$$

From examining Equations 2.6, 2.7, and 2.8, a way can be seen to improve the resolution of the optical microscopy or reduce the spot size of the focused light beam: use a shorter wavelength light or use an optical system with higher numerical aperture. For example, by using a blue laser of $\lambda = 400$ nm, NA = 0.85, a capacity of 25.3 GB/disk ($\Phi = 120$ mm) can be achieved in the DVR disk as compared with 0.65 GB/disk capacity in the CD-ROM of $\lambda = 780$ nm, NA = 0.47. There are, however, limitations, and it is difficult to break through the diffraction limit.

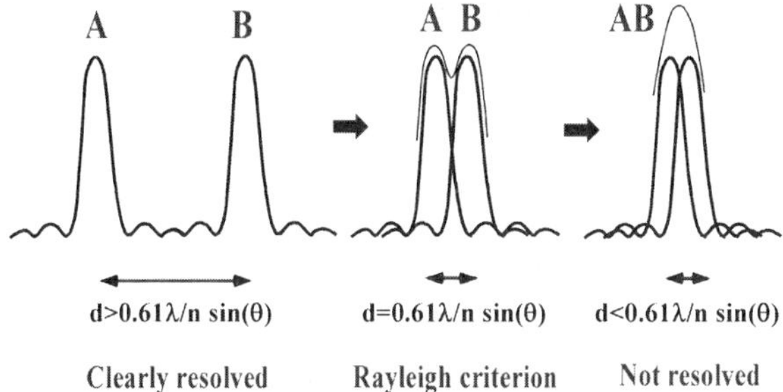

Figure 2.4 Limitation of the resolution of optical microscope due to the overlapping of the images when decreasing the distance of two points below the Rayleigh criterion.

A new kind of optical microscope, a laser scanning confocal microscope, has been developed to improve resolution and utilize the function of imaging in three dimensions (depth of field). In confocal microscopy, both the illumination and reflected laser beams are focused on the same spot on the surface or inside of the sample (see Figure 2.5). Normally, the sample is scanned and at every location of the light beam, the light intensity is detected and acquired by a computer. Because of the local illumination (the focused laser beam) and local detection, a resolution of $\sqrt{2}$ times higher than the resolution in an ordinary microscope can be achieved.[3] However, diffraction of laser light still limits the resolution.

The diffraction limitation of the light beam affects not only the resolving power of optical microscopy, but also all applications using far-field light such as optical spectroscopy, photolithography, and optical data storage. For some time, conventional optics has been limited by the wavelength of light due to the diffraction effect.

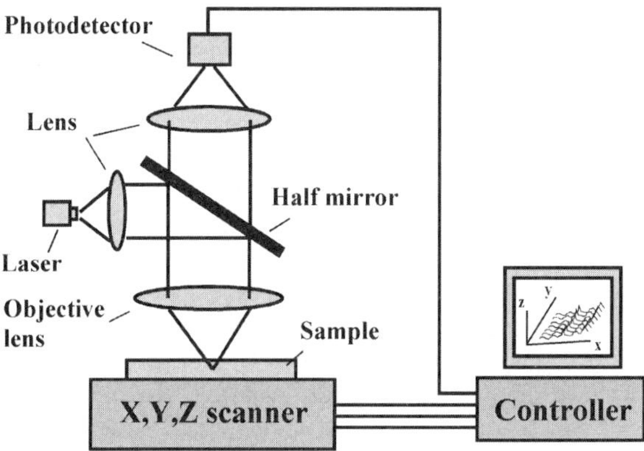

Figure 2.5 Simplified illustration of the laser confocal optical microscope.

2.2   *Concept of near-field optics and optical near-field microscopy*

The diffraction effect is not a barrier that is impossible to overcome. A new concept in near-field optics or near-field scanning optical microscopy breaks the wavelength barrier and moves into the subwavelength region. Near-field or evanescent-field are depicted in Figure 2.6. The concept of the near-field scanning optical microscopy (Figure 2.6a) had been proposed in 1928 by Synge.[4] He showed, for the first time, that the resolution of optical micros-copy could be below the wavelength if a sample is closely illuminated with a light source located near an aperture with subwavelength size, and the diffracted light is detected. His ideas were experimentally demonstrated by Ash in 1972 in the microwave region with centimeter radiation and resolu-tion of $\lambda/60$, far from the Abbe limitation.[5] After STM come about in 1982,[6] near-field optics and its related applications have become one of the most important topics of research with the pioneering works of Pohl et al.,[7] Lewis et al.,[8] Massey.[9]

The principle of near-field optics can be explained with the help of Fourier optics for a diffraction problem. It was shown that the light on the surface of an object to be imaged always diffracted into two components: (1) propagating wave with low spatial frequencies (corresponding to a lateral dimension that is larger than the Rayleigh criterion), and (2) evanescent wave with high spatial frequencies (corresponding to a lateral dimension that is smaller than the Rayleigh criterion). The evanescent wave is confined to a subwavelength dis-tance from the object corresponding to the near-field region. Since classical optics is concerned only with the far-field region, i.e., the propagating wave. Therefore, information of the high spatial frequency components of the dif-fracted wave (the near-field light) is lost in the far-field region.

Let's consider a sample in which the size or distance between the two components is finer than the wavelength of light. When illuminating the sample, for example, in total internal reflection, the evanescent field will be created and located near the fine structures (Figure 2.6b). Since the evanes-cent light does not propagate into the far-field, the fine structure is not observed in a conventional microscope with far-field optics. When the eva-nescent-field is disturbed by, for example, a subwavelength-sized scatterer that is placed closely to the field, the tip will interact with the evanescent-field and scatter photons (Figure 2.6c). As a result, the information of fine structures will be observed by detecting the scattered light in the far-field. This is the principle of the photon scanning tunneling microscope (PSTM) configuration that will be explained more in the next section.

The effect of converting the evanescent-field into the far-field by a scat-terer was experimentally observed by Newton. Newton's experiment showed that the evanescent-field on the surface of a prism at the total internal reflection will be converted into the far-field if a second prism is closely placed on the first prism's surface. When the gap between two prisms is

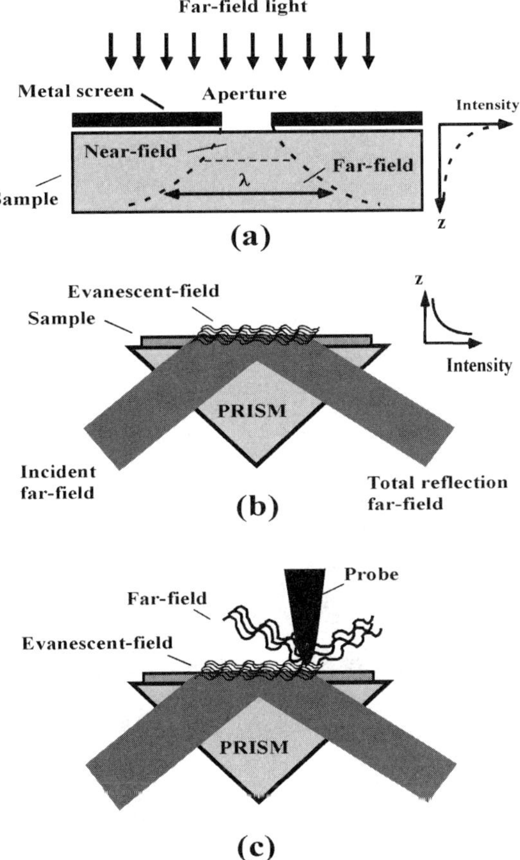

Figure 2.6 (a) Concept of the near-field optical microscopy proposed by Synge. Near-field is formed at the aperture on an infinite metal screen; (b) evanescent field on the surface of sample/prism where the total internal reflection is utilized; (c) utilizing the near-field information by approaching a small scatterer to disturb the evanescent field.

small enough (near-field interaction distance) the second prism will interact with the evanescent-field and emit it into the far field.

When a far-field light source is guided to a transmissive aperture of a size that is smaller than the optical wavelength, the light does not propagate but rather decays exponentially perpendicular to the aperture as a near-field light source. If this near-field light source is scanned in proximity to the sample, the near-field light is disturbed and scattered by the sample. Consequently, optical information of surface with resolution of about the aperture size can be observed in the far-field. In short, to utilize the conversion of near-field or evanescent-field light to the detectable far-field light, one must bring the scatterer or aperture close to the surface (near-field interaction distance).

Near-field light exists in a system where the light is uneasy to propagate such as ultra-small grating, small aperture (Figure 2.6a), optical waveguide, total reflection (Figure 2.6b), ultra-fine particles, etc. The essential features of the near-field light are (1) in contrast to far-field light, near-field light does not propagate, but is strongly localized; (2) optical near-field intensity (I) is damped out rapidly with increased distance (z) from the aperture as $I \sim z^{-4}$.[10] A finite difference time domain (FDTD) simulated distribution of the near-field light at the aperture of 300 nm diameter formed on a metallic screen is shown in Figure 2.7. A variation of the near-field intensity at the center of the aperture with the distance in z direction is shown in Figure 2.8. In these simulations (Figures 2.7, 2.8), a far-field light source of 780 nm wavelength was employed. The simulated results shown in Figures 2.7 and 2.8 have confirmed the localization characteristics of the near-field light at the sub-wavelength-sized aperture.

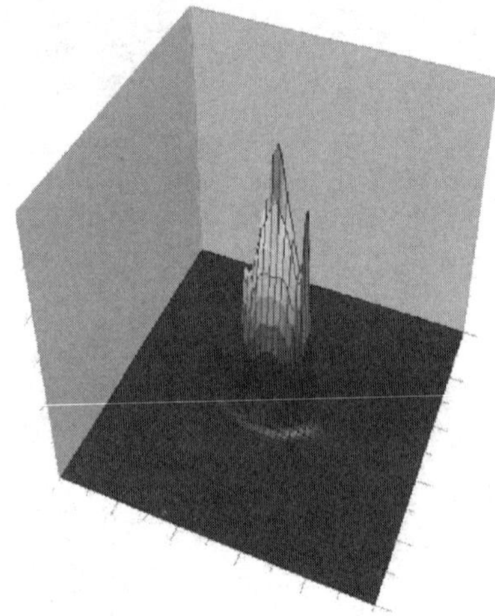

Figure 2.7 Distribution of near-field intensity at the aperture of 300 nm calculated by Finite Difference Time Domain (FDTD) simulation.

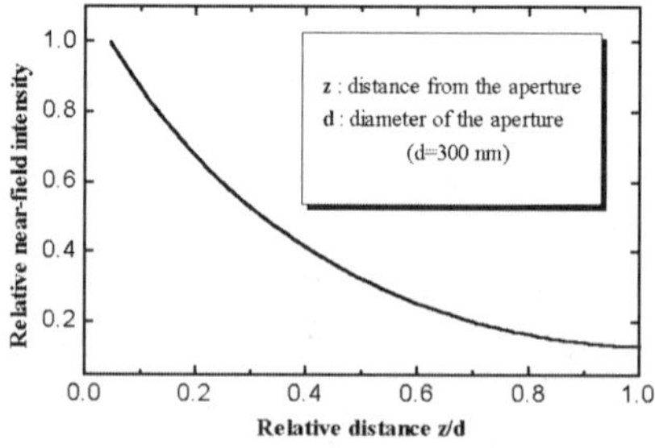

Figure 2.8 The variation of near-field intensity with the distance from the aperture according to the FDTD simulation. The data is taken at the center of the aperture shown in Figure 2.7.

Moreover, when the aperture becomes smaller and smaller, the near-field intensity drastically decreases. The optical throughput (the ratio of the near-field and corresponding far-field intensities) of a single hole with subwavelength diameter in an infinite metal film had been studied by Bethe using a scalar potential functional approach.[11] When the diameter of the hole (d) is smaller than the optical wavelength λ, the optical transmission scale as $(d/\lambda)^4$. The optical throughput of a single hole with different sizes on the infinite metallic screen and different wavelengths calculated with Bethe's model is plotted in Figure 2.9. It is seen that the optical throughput drastically decreases with decreasing of aperture diameter. To utilize near-field light with high resolution, a small, localized near-field light source is needed, as well as near-field intensity that is high enough to interact with the sample and to be detectable in the far-field. To look for a novel apertured tip with the enhancement of the optical throughput is very important for near-field applications.

To utilize the near-field light at the aperture (Figure 2.6a) or evanescent field on surfaces (Figure 2.6b), one must bring the near-field light source (aperture) or the scatterer close to the sample surface where the near-field is disturbed. The near-field probe is scanned over the sample or the sample is scanned under the probe in different modes of constant height, constant distance or constant photon intensity. However, as seen in Figure 2.8, the near-field intensity is drastically decreased with increasing distance between the aperture and the sample. Therefore, the constant distance mode is advantageous to avoid tip–sample damage and separate the topographical variation of the sample surface. That is why Synge's idea of subwavelength optical microscopy was delayed until the scanning technology was invented. After the invention of STM in 1982, the first near-field scanning optical microscopy (NSOM) or (SNOM) was experimentally invented.[7,8] It was confirmed that the optical resolution in NSOM is not determined by the wavelength of the light being used, but rather by the size of the aperture and the distance between the aperture and sample.

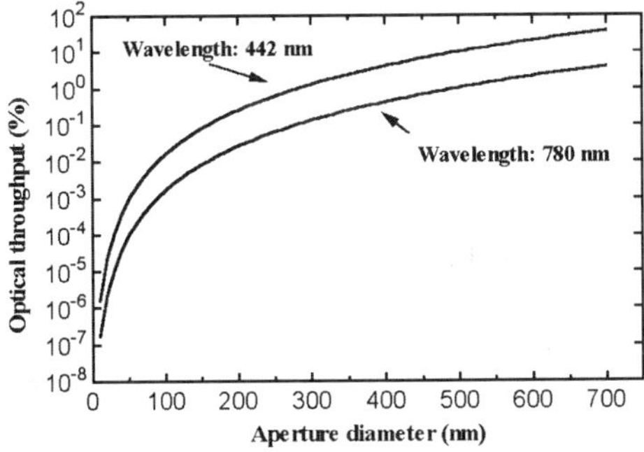

Figure 2.9 Optical throughput of near-field light emitted through an aperture on an infinite metal screen as a function of aperture diameter according to Bethe's model.

In practice, the aperture formed on a flat metallic screen (Figure 2.6a) as Synge's first proposal is not suitable for scanning. The optical near-field probe having a small aperture at the apex of a tapered, metallized quartz rod (Figure 2.10a), glass micropipette (Figure 2.10b), optical fiber (Figure 2.10c, 2.10d) or a micromachined cantilever (Figure 2.10e) is utilized. The main purpose of this work is to look at technological approaches for the fabrication of an apertured cantilever with very small aperture and high optical throughput.

In conventional optical microscopy, the image is observed by directly detecting the diffracted far-field light at the surface by an optical system; whereas in near-field microscopy, the image is observed by detecting the far-field light that results from the interaction of the probe and the diffracted near-field component on the surface. The contrast mechanism of the near-field imaging, therefore, strongly relates to the detail of the interaction between the tip and the sample. The resolution of the near-field scanning optical microscopy cannot be compared to the STM or AFM due to the limitation of the probe-sample interaction area (i.e., the aperture size and the "skin depth" effect of the coated metallic film[12] or the physical size of

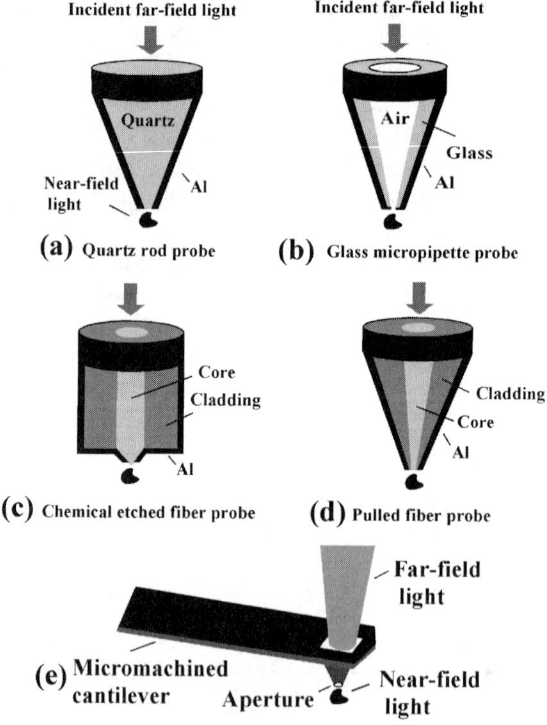

Figure 2.10 Schematic structure of actual apertured NSOM probes (a) metal-coated quartz rod; (b) metal-coated, taped glass micropipette; (c, d) metal-coated, taped optical fiber probe and (e) cantilevered probe with an aperture at the apex of the tip.

the scatterer). NSOM, however, combines the advantages of optical microscopy and spectroscopy with the high resolution of the SPM technique. Today,
NSOM has achieved spatial resolving power down to around 10 nm and
normally around 50–100 nm. That resolution is far from atomic level, but
NSOM has opened up a new frontier of nano-optics. Near-field optics is not
only applicable to subwavelength optical microscopy, but also for spectroscopy, subwavelength photolithography, next generation data storage and
many other applications in atomic and nano-optics and photofabrication.
NSOM may be utilized in air or in liquid that is a very powerful tool for
investigation (imaging and spectroscopy) of biological specimens. NSOM
has already been successful in imaging biological surfaces,[13] detecting single
molecules,[14,15] solid-state spectroscopy,[16] Raman spectroscopy, lithography
and data storage.[17-23]

2.3   *Instrumentation of optical near-field imaging*

There are two configurations in optical near-field imaging: near-field scanning optical microscopy (NSOM) that utilizes an ultra small aperture tip as
a near-field light source or detector (Figure 2.11); and photon scanning tunneling microscopy (PSTM) that utilizes a small dielectric or metallic tip as a
scattering center (Figure 2.12).

For the NSOM configuration, near-field imaging can be performed in
different modes of operation as follows: For transparent samples, transmission microscopy either in illumination (Figure 2.11a) or collection (Figure
2.11b) modes is used; for opaque samples, the NSOM image should be
observed in reflection mode (Figure 2.11c). A hybrid illumination/collection
mode is also effectively utilized in some cases (Figure 2.11d). For any kinds
of operation modes (shown in Figure 2.11), it is critical to utilize as small an
aperture as possible and to locate the aperture in proximity to the sample
to improve the resolution. The lateral resolution achieved with these setups
is typically on the order of 20–100 nm depending on the diameter of the
aperture and the distance between the aperture and the sample.

For the PSTM configuration, evanescent wave on the surface of the sample
is created by oblique far-field illumination using a prism in a total reflection.[24,25]
The sample is placed on the prism surface with an optical matching oil between
them. When the incident angle of the laser beam is larger than the critical
angle, the refracted wave on the surface of the sample shows an exponential
decay in the direction of the boundary surface. This is referred to the *evanescent
wave* and would follow the spatial variation of the sample surface. By bringing
a small dielectric tip toward the surface, at a distance close to the penetration
depth of the effective field, the tip acts as a scattering center and a coupler to
convert the evanescent field into a propagating wave that can be detected in
the far-field by a photodetector (Figure 2.12a). With this configuration, an
evanescent-field photon is likely to tunnel from the sample surface to the
dielectric tip (the same as electron tunneling in the STM), hence the name

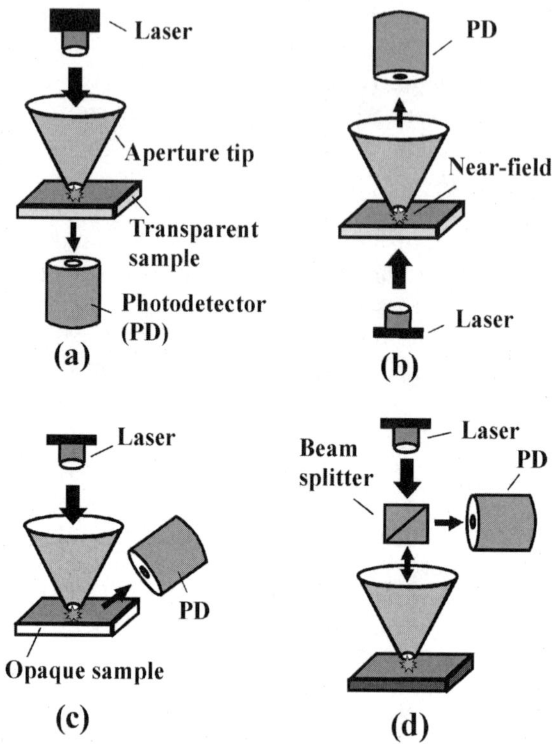

Figure 2.11 Operation modes for optical near-field imaging with NSOM configuration (a) illumination mode with an aperture probe in transmission; (b) collection mode with an aperture probe in transmission; (c) reflection mode with and aperture probe on an opaque sample; (d) hybrid illumination-collection mode.

photon scanning tunneling microscope (PSTM). The scattered light contains information of the sample features. When the size of the probe and the distance from the probe to the sample is sufficiently small compared to the wavelength, the near-field process is scaled, not by the wavelength but by the size of the system. Commonly, a spatial resolution of 10–100 nm is typically achieved with the PSTM configuration.

In order to enhance the scattering cross-section of the tip, a metallic tip is used instead of a dielectric one and the scattered light is detected by a photodetector placed behind the tip[26-29] (Figure 2.12b). In some cases, a resolution down to 1 nm was reported.[26] To enhance the scattering efficiency, a scanning plasmon near-field microscope (SPNM) was proposed. In the SPNM configuration, surface plasmon generated at the surface of an ultra-thin metallic film on the prism and a metallic coated tip is utilized. A spatial optical resolution of 3 nm was achieved.[30] Another idea of PSTM was proposed in which a small photodetector integrated at the end of the probe was used to detect the near-field light on the surfaces[31,32] (Figure 2.12c). An optical resolution of 20 nm was achieved with this probe.[31] In some cases, to reduce

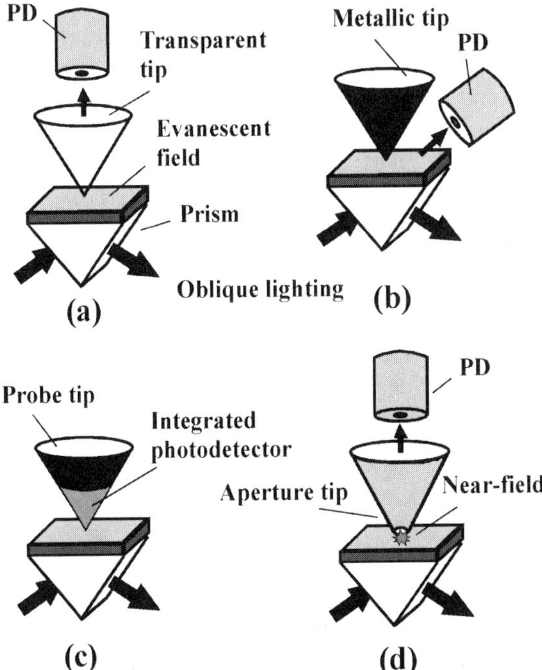

Figure 2.12 Operation modes for optical near-field imaging with PSTM configuration (a) PSTM configuration with a transparent tip and oblique illumination; (b) enhanced scattering cross section with a metallic tip and oblique lighting; (c) PSTM configuration with an integrated detector at the tip; (d) PSTM configuration with an apertured probe.

a large background signal on the surface, the apertured probe is used in the PSTM configuration as shown in Figure 2.12d.

Compared to the NSOM, the PSTM configuration is easier to operate and it is easier to fabricate the probe, but suffers somewhat from difficulties in data interpretation. It is difficult to extract the near-field signal from a relatively huge background signal caused by the scattering of the sample. The nature of the optical signal in the PSTM configuration is still under consideration. The NSOM configuration is more useful compared to the PSTM, but it does suffer from the technological problems in fabrication of the apertured tip. This work mainly deals with finding a technological approach for fabrication of the cantilevered NSOM probe having a small aperture and developing the applications with the fabricated apertured probes.

2.4 *Techniques for control of the tip–sample distance*

As previously mentioned, to utilize the near-field light generated at the aperture (NSOM configuration) or evanescent field generated on the surface of the sample (PSTM configuration), the probe must be maintained at a constant proximity to the surface of the sample. There were several techniques for the regulation of the tip–sample distance as the following (see also Figure 2.13).

- Constant-height mode: the probe is scanned at a fixed height parallel to the mean height of the sample. No feedback is required in this mode. As seen in Figure 2.13a, the probe is scanned following a straight line above the surface.
- Constant-distance mode: the tip is scanned following the topography of the sample using auxiliary feedback, such as a tunneling current,[7,12,33,34] atomic force, or shear force.[35,36] As shown in Figure 2.13b, the probe is scanned following the topography of the surface.
- Constant-intensity mode: the detected optical signal is kept constant during scanning by varying the tip–sample distance using a feedback system with the detected optical signal.[37-39] As seen in Figure 2.13c, the probe is scanned following the optical near-field contour on the surface.

The constant-intensity mode is not very useful because both the topography and the transmissivity of the sample influence the near-field signal. Moreover, the lack of optical signal in opaque parts of the sample would

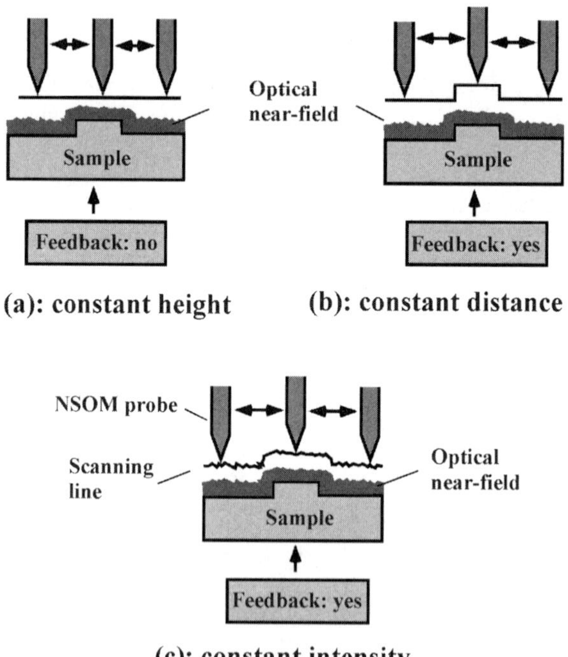

Figure 2.13 Scanning modes for near-field imaging (a) constant height mode; (b) constant distant mode; (c) constant intensity mode.

result in a crash of the tip and the sample. The constant-height mode can be used for the case of very smooth sample. Nevertheless, it is difficult to avoid damage to the probe and sample. The constant-distance mode is the most widely used in NSOM applications.

There are several methods to control the tip–sample distance in the constant-distance operation mode. For the conductive tip and sample, the gap can be kept constant by utilizing the tunnel current (or STM principle); see Figure 2.14a).[7,12,33,34] A tunneling current is used between a protrusion of the apertured metal tip and the conductive sample that exponentially depends on the tip–sample distance being kept constant at several nA. A feedback loop circuit is used to control the voltage applied to a piezo element, either on the tip or on the sample, so that the tip–sample distance is regulated according to the topography of the surface. At every point of scanning area an optical signal is recorded. The near-field image is reproduced by drawing the optical signal as a function of the corresponding X, Y positions. This technique is applicable only for a conductive sample.

For optical fiber or glass micropipette probes, the so-called shear-force technique has been widely used (see Figure 2.14b). The shear-force technique was first proposed in 1992 by Toledo-Crow et al.[35] and Betzig et al.[36] The probe is mounted on a piezoelectric or tuning fork element. The tip is vibrated in a motion parallel to the sample surface at its mechanical resonance. When the tip approaches the sample to around ten nanometers, the shear-force between the tip and the sample is activated resulting in a decrease of the vibration

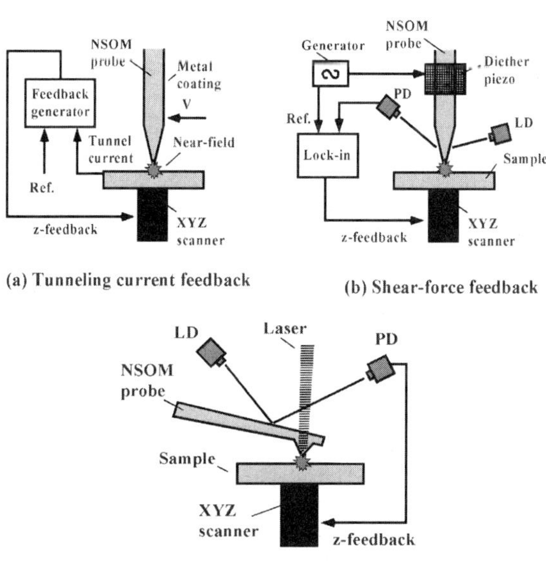

Figure 2.14 Regulation of the tip–sample distance by (a) tunneling current; (b) shear force techniques; (c) atomic force.

amplitude. By keeping the vibration amplitude at a constant value, the tip-sample distance can be fixed constant by modulating the voltage applied to the piezo element. The amplitude of the tip vibration is monitored by, for example, optical differential interferometer or by measuring the oscillation in the reflected laser beam focused at the tip end. The optical signal is extracted using a lock-in amplifier technique. The shear-force feedback mechanism is widely used in most NSOM systems with optical fiber probes. This provides a means of avoiding a crash between the tip and the sample, and enables the tip to be kept at a short distance during the scan (generally estimated to be less than 10 nm). Since the fiber tip is laterally vibrated, the optical resolution is also limited by the vibration amplitude of the tip. The exact mechanism of the shear force and the relationship between constant-distance regulation and optical properties of the surface is not yet well understood.

In the case of a cantilever type, a more convenient way to keep a constant gap during scanning is to keep the atomic force between the tip and sample at a constant value (see Figure 2.14c). The atomic force between the tip and the sample causes cantilever bending. The bending of the cantilever is monitored and a feedback loop used to control the voltage applied on the piezo on the probe or sample. This regulation technique is effectively used because of the good understanding of atomic force microscopy and force interactions are not limited with respect to the tip and sample material. To utilize the cantilever type probe with the AFM regulation technique, the cantilever should have suitable mechanical properties.

As previously mentioned, conventional optical microscopy is advantageous for various contrast mechanisms. However, in near-field imaging, the probe-sample interaction can make the contrast more complicated because many effects can alter the contrast mechanism, such as details of the probe field, interaction between the probe and sample.[40-43] A complete understanding of the contrast mechanism in near-field optics has not been achieved. Research has shown that near-field images might be affected by topography and the dielectric contrast of the sample. The motion of the tip or the sample may produce an artifact. Even under certain conditions, the contrast of the image may be dominated by the artifact contribution.[44] That is why near-field optics and its related technologies still need further development. There is room for further study both in the experimental and theoretical aspects.

2.5 *Tapered optical fiber-based near-field optical probes*

There are at least three critical parts in the NSOM probes: the tapered region, i.e. the tip; the optical aperture; and the metallic opaque screen around the optical aperture. Optical performance of the NSOM probe strongly depends on the size and shape of the tapered region and the optical aperture. Fabrication of the NSOM probe with high reproducibility, in nanometric scale, of the shape and the size of the tapered region, as well as the aperture, is the most important task. The technique of regulation of the tip–sample distance and method of detection is also essential in optical near-field applications.

At the first stage of the NSOM, a tapered quartz rod (Figure 2.10a) was used. A quartz rod is manually cut and polished to a pointed shape and then covered with a metallic film. A small aperture was formed by pressing the tip against the sample surface until a tiny spot of light became visible under an optical microscope. An aperture at least 20 nm in diameter can be created by this technique.[7] This method, however, is not very reproducible, has low production yield, and the aperture becomes larger after several hours of operation due to the thermal effect.[7,10] Another method was proposed by Lewis et al.,[8,45-47] in which the aperture is formed at the end of a metal-coated tapered glass micropipette (Figure 2.10b) by heating and pulling the glass capillary tube. A metal film covers the outer side of the pipette leaving an aperture of smaller than 100 nm.

Presently with the development of optical fiber technology, most of the NSOM probes are made from optical fiber (Figures 2.10c and 2.10d). Details of the fabrication process and techniques for improving the performance of the optical fiber probes can be found in references.[48-55] There are two basic techniques to produce the tapered region in the optical fiber: a selective chemical etching[49-53] or a thermal heating and pulling,[54,55] and in some cases both of them are used.

In the thermal heating and pulling method, optical fiber is mounted on a puller as shown in Figure 2.15. A high-power laser (typically a CO_2 laser) is used to locally heat the fiber and the puller is mechanically adjusted. The heated part of the fiber is melted, gradually tapered and automatically broken by adjusting the puller, leaving a conical tapered tip. The tapered angle and overall geometric structure of the pulling tip is dependent on the power of the laser, time of heating and pulling force. In some advanced systems, the pulling process is monitored and controlled by a computer. With this method, it has been possible to fabricate a tapered fiber with an apex diameter of 50 nm and tapered angle in the range of 20–40°. However, it is difficult to control the tapered angle and apex diameter. Generally, the reproducibility of the probe using this method is quite low. Optical images of an optical fiber based probe fabricated by the heating and pulling method are shown in Figure 2.16. The weak near-field light emitted from the aperture is visible in Figure 2.16b. The tapered region of the probe is quite long and most of the light is lost in that region, consequently, the optical throughput is low.

In the chemical etching method, a chemical etchant of $NH_4F:HF:H_2O$ (buffered–HF) is used to etch the fiber. A minimum opening angle of the tip (θ) smaller than 15°, and a tip as sharp as 1 nm can be achieved with this method. The θ angle can be varied by precisely controlling the composition of the buffered–HF (BHF) solution as well as the composition of the optical fiber core or cladding. The technique of chemical etching was systematically developed by Ohtsu et al., and the improvements achieved with this technique are reported in detail in that monograph.[49]

Advantages of the chemical etching method in comparison to the thermal heating and pulling method are the capability to control the shape, the opening angle and the size of the tapered region. Several novel structures

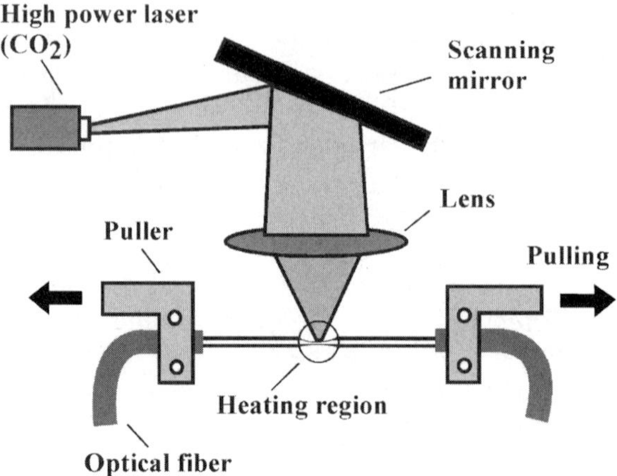

Figure 2.15 Illustration of the thermal heating and pulling method.

with double or triple tapered angles that improve the optical throughput
have been developed. The chemical etching method can make the tip with
higher reproducibility compared to the heating and pulling method. However, the chemical etching technique still suffers from high production yield
and improvement of the flatness of the tapered region is needed.

Experimental and theoretical works have reported[54,56-58] that the cone
angle of the tapered region and the overall aspect ratio of the tip are very
important and directly determine the optical throughput of the NSOM probe.
The fact that the chemical etching method can more precisely control the
shape and tapered angle of the optical fiber NSOM tip is clear.

Another critical step in the fabrication of optical fiber NSOM probes is
proper formation of the aperture and opaque metallic screen around the
aperture. After producing the tapered fiber tip by either of the above methods, an optical aperture is then created by appropriate metallization and
selective etching with the process shown in Figure 2.17. First, the exterior of
the tapered optical fiber tip is coated with an opaque metallic film, such as
aluminum (Al) or gold (Au), with sufficient thickness to avoid the illumination of excitation light or the detection of the background signal through the
side wall of the tapered region. Oblique metal evaporation on the rotary
fiber is typically used (Figure 2.17a). Next, the metal at the end of the tip is
selectively removed by wet etching with a suitable metal etchant (Figure
2.17c) or milling with a focused ion beam (Figure 2.17d) to form an optical
aperture (as shown in Figure 2.17e).

The size and shape of the aperture of the fiber tip that is formed is not
consistently reproducible. The technique also suffers from high production
yield. Optical fiber NSOM probes are regarded as a metal-cladding waveguide.
Since the metallic waveguide has a complex loss mechanism due to the absorption and the existence of "cut-off effect," the optical throughput of the fiber
probe is normally very low (about 10^{-6}–10^{-5} for 100 nm apertures).[54] For

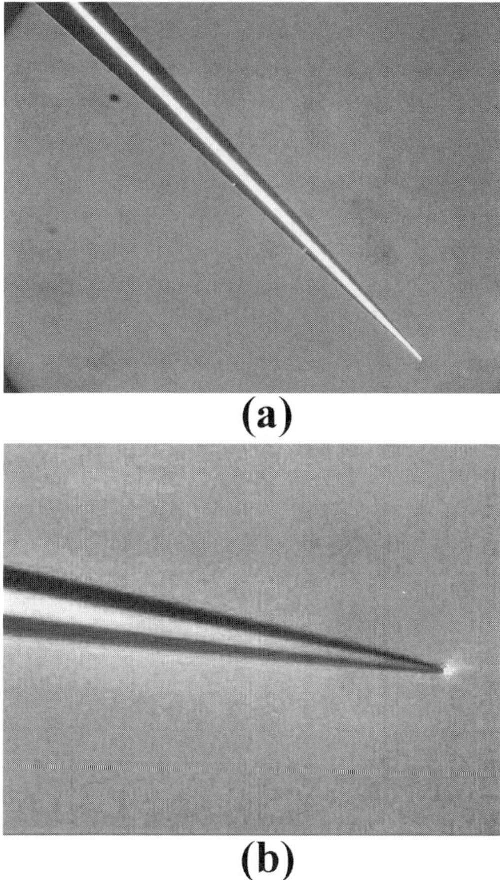

Figure 2.16 Optical images of a fiber probe made by heating and pulling method (a) without light guiding; (b) with light guiding.

sub-100 nm apertures, only HE_{11} mode with a low cut-off diameter can come close to the aperture, whereas most other modes are cut off at the region of the fiber where the diameter is smaller than the wavelength.[59] These cut-off modes are absorbed by the metallic waveguide, which can lead to thermal damage of the probe.[60-61] Furthermore, the optical fiber tip is very fragile and can be easily broken at a single touch of the sample. To utilize an NSOM probe with a fiber tip, one must keep the aperture at a constant distance to the sample by the shear-force technique, which results in the NSOM system being complicated and limits the lateral optical resolution of the probe.

2.6   Advantages of Si micromachined probes

Optical fiber based NSOM probes play an important role in the development of near-field optics. Their most important advantages are the convenience

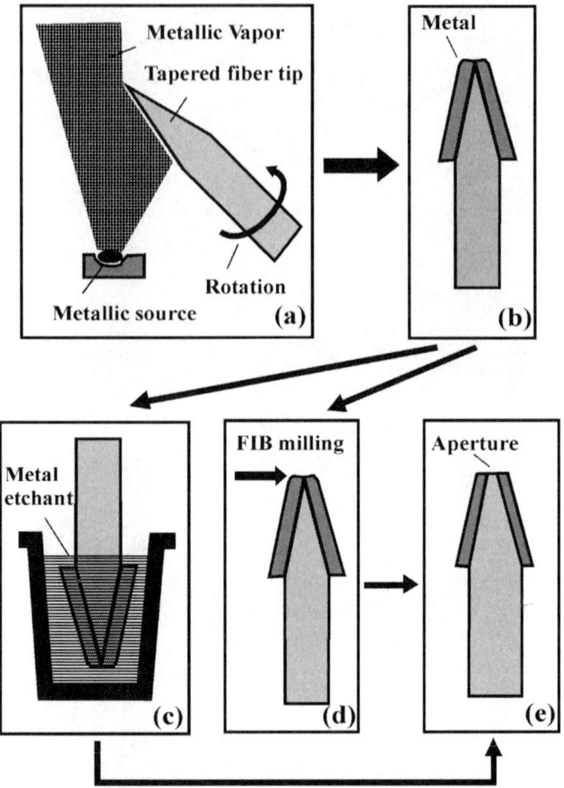

Figure 2.17 Techniques for forming optical aperture at the apex of the tapered fiber tip (a) rotary metal evaporation; (b) metal-coated fiber tip (c) selective metal etching to form aperture (d) focused ion beam etching to form aperture (e) fiber probe with the aperture after doing step c or step d.

in coupling the far-field light from a light source to the aperture or from the aperture to the detectors. The optical fiber itself is inexpensive and popular. Although performance of optical fiber based NSOM probes has been improving and they are commercially available, problems remain:

- Generally the process from optical fiber to the optical fiber NSOM probe is not suited for reproducible mass production. These probes are still produced one by one and it is an expensive process, although the optical fiber itself is inexpensive.
- The current techniques used for forming the tip and optical aperture make it hard to control the shape and nanoscale sized accuracy.
- The optical throughput of the conventional fiber NSOM probes is very low due to a strong absorption of light at the metallic waveguide and the cut-off effect. Only a small fraction of light can pass through

the tapered region and go toward the aperture of the probe. Several advanced fiber NSOM probes with higher optical throughput have been reported, but they still suffer from high production yield and are not commercially produced yet.

- Since the optical throughput is low, most of the incident light is either absorbed by the metal coating or reflected back. If the input power is increased, heat may modify the sample or enlarge the aperture.
- The optical fiber tip is very fragile and easily broken when touching the surface of a sample that has a high surface topography.
- The tip–sample distance is regulated by shear-force technique, which makes the system more complicated and limits the lateral optical resolution.
- It is difficult to utilize an optical fiber probe in several applications that require an array of nanolight source, for instance, in near-field optical data storage or multiprocessing using near-field light.

Presently the near-field optic community relies heavily on optical fiber based probes because there currently are no other commercially produced probes available. One of the main reasons that NSOM has not yet become a widely used technique is the lack of high quality, reproducibility and inexpensive probes. Tip and aperture fabrication remains a major problem. The findings presented in this work are new technological approaches to aperture fabrication of NSOM probes using Si micromachining technology. Current drawbacks of the optical fiber based NSOM technique are, for the most part, overcome. The advantages of the Si micromachined probes and a comparison of the micromachined and conventional fiber probes is shown in Table 2.1.

Table 2.1 Comparison of the Conventional Optical Fiber and Micromachined NSOM Probes

Comparison parameters	Optical fiber probe	Silicon micromachined probe
Fabrication	No mass production (expensive)	Mass production (less expensive)
Reproducibility	Poor	Higher
Optical throughput	Low (10^{-6}–10^{-3} for 100-nm aperture)	Higher (10^{-3}–10^{-2} for 100-nm aperture)
Polarization control	Difficult	Possible
Mechanical properties	Poor	Good
Aperture size	20–200 nm (mostly)	20–200 nm (mostly)
Measurement system	NSOM system	AFM system available
Tip–sample distance	By shear force (more complicated)	By AFM force (well understood)
Probe array capability	Difficult	Easy
Non-optical applications	Difficult	Possible

References

1. Born, M. and Wolf, E., *Principles of Optics*, 6th ed., Pergamon Press, Oxford, 1980.
2. Hecht, E., *Optics*, Addison-Wesley, Reading, MA; Longman, Chicago, 1998, p. 37.
3. Courjon, D. et al., Near-field optical microscopy: fifteen years of existence, *Condensed Matter News*, 6(3–4), 14, 1998.
4. Synge, E.H., A suggested model for extending microscopic resolution into the ultra-microscopic region, *Phil. Mag.*, 6, 356, 1928.
5. Ash, E.A. and Nichols, G., Super-resolution aperture scanning microscope, *Nature*, 237, 510, 1972.
6. Binnig, G. and Rohrer, H., Scanning tunneling microscopy, *Helv. Phys. Acta*, 55, 726, 1982.
7. Pohl, D.W., Denk, W., and Lanz, M., Optical stethoscopy: image recording with resolution $\lambda/20$, *Appl. Phys. Lett.*, 44, 651, 1984.
8. Harootunian, A. et al., Super resolution fluorescence near-field scanning optical microscopy, *Appl. Phys. Lett.*, 49, 674, 1986.
9. Massey, G.A., Microscopy and pattern generation with scanned evanescent waves, *Applied Optics*, 23, 658, 1984.
10. Pohl, D.W., in *Scanning Tunneling Microscopy II*, Wiesendanger, R., Guntherodt, H.J., Eds., Springer-Verlag, Heidelberg, Germany, 1995, p. 233.
11. Bethe, H.A., Theory of diffraction by small holes, *Phys. Rev.*, 66, 163, 1944.
12. Duerig, U., Pohl, D.W., and Rohner, F., Near-field optical scanning microscopy, *J. Appl. Phys.*, 59, 3318, 1986.
13. Betzig, E. et al., Near-field fluorescence imaging of cytoskeletal actin, *Bioimaging*, 1, 129, 1993.
14. Betzig, E. and Chichester, R.J., Single molecules observed by near-field scanning optical microscopy, *Science*, 262, 1422, 1993.
15. Ambrose W.P. et al., Single molecule detection and photochemistry on a surface using near-field optical exitation, *Phys. Rev. Lett.*, 72, 160, 1994.
16. Grober, R.D. et al., Optical spectroscopy of a GaAs/AlGaAs quantum wire structure using near-field scanning optical microscopy, *Appl. Phys. Lett.*, 64, 1421, 1994.
17. Ono, T. and Esashi, M., Subwavelength pattern transfer by near-field photolithography, *Jpn. J. Appl. Phys.*, 37, 6745, 1998.
18. Tanaka, S. et al., Printing sub-100 nanometer features near-field photolithography, *Jpn. J. Appl. Phys.*, 37, 6739, 1998.
19. Wegscheider, S. et al., Scanning near-field optical lithography, *Thin Solid Films*, 264, 264, 1995.
20. Ghislain, L.P. et al., Near-field photolithography with a solid immersion lens, *Appl. Phys. Lett.*, 74, 501, 1999.
21. Rogers, J.A. et al., Generating ~90 nanometer features using near-field contact mode photolithography with an elastomeric phase mask, *J. Vac. Sci. Technol.*, B16, 59, 1998.
22. Betzig, E. et al., Near-field magneto-optics and high density data storage, *Appl. Phys. Lett.*, 61, 142, 1992.
23. Hosaka, S. et al., SPM-based data storage for ultrahigh density recording, *Nanotechnology*, 8, 58, 1997.

24. Fornel, F. et al., An evanescent field optical microscope, *Proc. SPIE*, 1139, 77, 1989.

25. Courjon, D., Sarayeddine, K., and Spajer, M., Scanning tunneling optical microscopy, *Opt. Commun.*, 71, 23, 1989.

26. Zenhausern, F., O'Boyle, M.P., and Wickramasinghe, H.K., Apertureless near-field optical microscopy, *Appl. Phys. Lett.*, 65, 1623, 1994.

27. Zenhausern, F., Martin, Y., and Wickramasinghe, H.K., Optical imaging at 10 angstrom resolution, *Science*, 269, 1083, 1995.

28. Inoue Y. and Kawata, S., Near-field optical microscopy using a metallic vibrating tip, *Opt. Lett.*, 19, 159, 1994.

29. Inoue Y. and Kawata, S., A scanning near-field optical microscope having scanning electron tunneling microscope capability using a single metallic probe tip, *J. Microsc.*, 178, 14, 1995.

30. Specht, M., Scanning plasmon near-field microscope, *Phys. Rev. Lett.*, 68, 476, 1992.

31. Akamine, S., Kuwano, H., and Yamada, H., Scanning optical microscope using an atomic force microscope cantilever with integrated photodiode, *Appl. Phys. Lett.*, 68, 579, 1996.

32. Davis, R.C., Williams, C.C., and Neuzill, P., Micromachined submicrometer photodiode for scanning probe microscopy, *Appl. Phys. Lett.*, 66, 2309, 1995.

33. Pohl, D.W., Denk, W., and Durig, U., Optical stethoscopy: imaging with $\lambda/20$, *Proc. SPIE*, 565, 56, 1985.

34. Betzig, E., Isaacson, M., and Lewis, A., Collection mode near-field scanning optical microscopy, *Appl. Phys. Lett.*, 51, 2088, 1987.

35. Toledo-Crow, R. et al., Near-field differential scanning optical microscope with atomic force regulation, *Appl. Phys. Lett.*, 60, 2957, 1992.

36. Betzig, E., Finn, P.L., and Weiner, J.S., Combined shear force and near-field scanning optical microscopy, *Appl. Phys. Lett.*, 60, 2484, 1992.

37. Courjon, D., Sarayeddine, K., and Spajer, M., Scanning tunneling optical microscopy, *Optics Comm.*, 71, 23, 1989.

38. Vigoureux, J.M., Girard, C., and Courjon, D., General principles of scanning tunneling optical microscopy, *Opt. Lett.*, 14, 1039, 1989.

39. Reddick, R.C., Warmack, R.J., and Ferrell, T.L., New form of scanning optical microscopy, *Phys. Rev. B*, 39, 767, 1989.

40. Valle, P.J., Greffet, J.J., and Carminati, R., Optical contrast, topographic contrast and artifacts in illumination-mode scanning near-field optical microscopy, *J. Appl. Phys.*, 86, 648, 1999.

41. Vigoureux, J.M., Depasse, F., and Girard, C., Superresolution of near-field optical microscopy defined from properties of confined electromagnetic waves, *Appl. Opt.*, 31, 3036, 1992.

42. Vigoureux, J.M. and Courjon, D., Detection of nonradiative fields of light in the Heisenberg uncertainty principle and the Rayleigh criterion, *Appl. Opt.*, 31, 3170, 1992.

43. Hulst, N.F. et al., Evanescent field optical microscopy: effect of polarization, tip shape and radiative waves, *Ultramicroscopy*, 42–22, 416, 1992.

44. Hecht, B. et al., Facts and artifacts in near-field optical microscopy, *J. Appl. Phys.*, 81, 2492, 1997.

45. Lieberman, K. and Lewis, A., Simultaneous scanning tunneling and optical near-field imaging with a micropipette, *Appl. Phys. Lett.*, 62, 1335, 1993.

46. Lewis A. and Lieberman, K., Near-field optical imaging with a nonevanescently excited high-brightness light source of subwavelength dimensions, *Nature*, 354, 214, 1991.

47. Betzig, E., Isaacson, M., and Lewis, A., Collection mode near-field scanning optical microscopy, *Appl. Phys. Lett.*, 51, 2088, 1987.

48. Betzig, E. et al., Breaking the diffraction barrier: optical microscopy on a nanometric scale, *Science*, 251, 1468, 1991.

49. Ohtsu, M., *Near-field Nano/Atom Optics and Technology*, Springer-Verlag, Tokyo, 1998.

50. Hoffmann, P., Dutoit, B., and Salathe, R.P., Comparison of mechanically drawn and protection layer chemically etched optical fiber tips, *Ultramicroscopy*, 61, 165, 1995.

51. Pangaribuan, T. et al., Reproducible fabrication technique of nanometric tip diameter fiber probe for photon scanning tunneling microscope, *Jpn. J. of Appl. Phys.*, 31, L1302, 1992.

52. Zeisel, D. et al., Pulsed laser-induced deposition and optical imaging on a nanometer scale with scanning near-field microscopy using chemically etched fiber tips, *Appl. Phys. Lett.*, 68, 2491, 1996.

53. Ohtsu, M. et al., Nanometer resolution photon STM and single atom manipulation, in *Near Field Optics*, Pohl, D.W. and Courjon, D., Eds., NATO ASI series (242), Kluwer, Dordrecht, 1993, p. 131.

54. Valaskovic, G.A., Holton, M., and Mirrison, G.H., Parameter control, characterization, and optimization of optical fiber near-field probes, *Appl. Opt.*, 34, 1215, 1995.

55. Fischer, U.Ch. and Zapletal, M., The concept of a coaxial tip as a probe for scanning near-field optical microscopy and steps towards a realization, *Ultramicroscopy*, 42–44, 393, 1992.

56. Betzig, E. et al., Breaking the diffraction barrier: optical microscopy on a nanometric scale, *Science*, 251, 1468, 1991.

57. Novotny, L., Pohl, D.W., and Hecht, B., Scanning near-field optical probe with ultra small spot size, *Opt. Lett.*, 20, 970, 1995.

58. Pohl, D.W. et al., Radiation coupling and image formation in scanning near-field optical microscopy, *Thin Solid Films*, 273, 161, 1996.

59. Novotny, L. and Hafner, C., Light propagation in a cylindrical waveguide with a complex, metallic dielectric function, *Phys. Rev. E*, 50, 4094, 1994.

60. Yakobson, B.I. et al., Thermal/optical effects in NSOM probes, *Ultramicroscopy*, 61, 179, 1995.

61. Staehelin, M. et al., Temperature profile of fiber tips used in scanning near-field optical microscopy, *Appl. Phys. Lett.*, 68, 2603, 1996.

chapter three

Introduction to silicon micromachining technology

Silicon microfabrication is the most appropriate technology today used to fabricate micro sensors, actuators and microsystems for applications in industries such as automotive, aircraft, display, communications, robotics, and disposable medical products. Silicon microfabrication is also an appropriate technology for the fabrication of microscanning probes with well-defined properties and other tools for touching the "nanoworld." Advantages of Si micromachining technology include miniaturization, precisely controllability, multi-integration and functionality, high reproducibility, mass production and low cost. New devices based on Si micromachining technology are being used in many commercial areas. As mentioned in Chapter 2, the fabrication of optical near-field probes require precisely controlled nanometric scales. Silicon micromachining techniques can be suitable for fabrication of NSOM probes.

There are a number of good references in the field of Si micromachining.[1-5] Appropriate information regarding the fundamentals, principles and applications of Si micromachining technology is available in these comprehensive books. Fundamental techniques of Si micromachining that were used for the fabrication of the near-field scanning optical microscopy (NSOM) and other probes are described here.

3.1 Lithography

3.1.1 Photolithography

Photolithography is a technique used in microelectronics and micromachining. Photolithography is used for transfering very small and well-defined patterns from a designed mask to a thin film of photoresist onto a substrate. The photoresist pattern is used as a mask for the next processing step on the wafer, such as selective etching or selective deposition.

The principle of conventional photolithography is easily understood with an example on SiO_2/Si substrate using an immersion mask shown in Figure 3.1. The desired patterns are designed and formed either on the

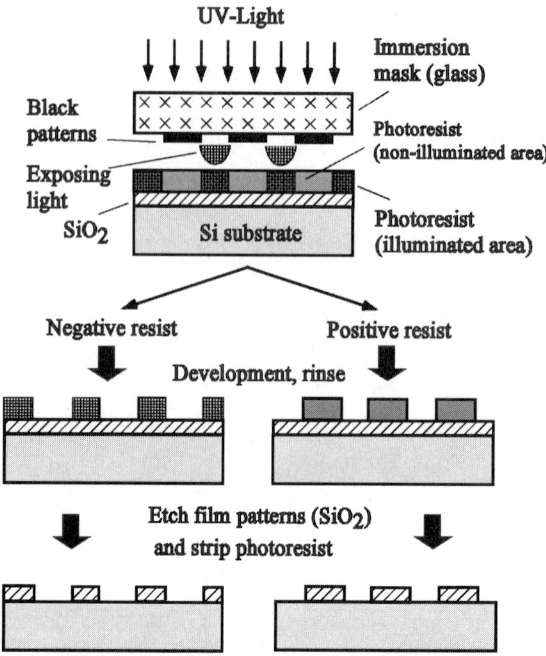

Figure 3.1 Principles of photolithography.

immersion or a Cr mask with "black" and "white" patterns. The mask is aligned and contacted with a photoresist coated substrate with an aligner. An ultraviolet (UV) light, such as a mercury lamp, is used to expose to the mask. The UV light passing through the "white" pattern on the mask will induce a chemical reaction in the desired area of the photoresist layer, altering the solubility of the resist in a solvent. There are two kinds of photoresist, positive and negative. Positive means degradation of the resist material (polymer) and negative means cross-linking of the polymer. After exposure and development in the solvent, only the desired resist patterns exist on the surface. The resist patterns usually serve as a mask layer for the next etching step, for example, SiO_2 etching, lift-off, metal etching and glass etching. Since microelectronics, Si micromaching and photolithography deal with the "microworld" or even the nanoworld, a clean room is preferable during the process.

Normally, the resolution of photolithography (i.e., the smallest patterns or distance between two nearest patterns which can be created on the photoresist layer after the photolithography process) is limited by the wavelength of the exposed light and the amount of contact between the mask and the photoresist layer. Usually, the immersion mask is used for forming patterns with sizes larger than several μm. For forming smaller patterns, the Cr mask is preferable. In this work, both positive and negative photolithography for patterning the Si and glass substrate was employed. The sequence of the Si cleaning and photolithography process is expressed as follows:

- Sample cleaning: standard RCA method
- Baking: at 145°C, 30 min for removing the adsorbed water from the surface
- Photoresist spin coating (negative resist OMR-83 [Tokyo Oukakogyou Co.]: 60 cp, spinning speed: 3000 rpm, spinning time: 30 seconds, thickness of photoresist after spinning: around 2 µm. Positive resist OFPR-800 [Tokyo Oukakogyou Co.]: 200 cp, spinning speed: 3000 rpm, spinning time: 30 seconds, thickness of photoresist after spinning: around 3 µm)
- Prebaking: (negative resist: baking at 75°C for 30 min.; positive resist: baking at 90°C for 30 min.)
- Expose: a mercury lamp, 2 mW/cm^2 was used (for negative resist exposing time is 18 seconds; for positive resist exposing time is 150 seconds)
- Development: negative resist is developed in an OMR developer at room temperature for 180 seconds and rinse in Trichloroethane: Ethanol =1:1 for 60 seconds; positive resist is developed in NMD-3 developer, at room temperature in 150 seconds and rinse with DI-water for 60 seconds
- Postbaking: 145°C, 20–30 minutes to remove residual developing solvent and anneal the film to promote interfacial adhesive and enhance the hardness of the photoresist film
- Oxygen plasma etching: 0.5–1.5 min., RF power of 100 W for removing unwanted resist left behind after development

3.1.2   Electron beam lithography

In specialized applications, in which photolithography cannot be used, electron beam (EB) lithography is used to make nanoscale-sized patterns or patterns with nanometric precision in size. The EB lithography can either be used for directly writing patterns on the substrate or forming the Cr mask for photolithography. In EB lithography, the desired patterns are prepared by a computer program. The computer-stored patterns are directly converted to address the electron beam exposing on the sample. A simplified illustration of the EB lithography system is shown in Figure 3.2. An electron beam emitted from a thermal or a field emission emitter is accelerated and focused with electron optics systems in a high-vacuum chamber. The focused electron beam is used for writing the pattern on the EB resist. In contrast to photolithography, the resolution of the EB lithography is not determined by the wavelength of the electron beam, but rather by the scattering of the electron beam inside the resist and the back scattering of the electrons from the substrate to the resist. Depending on the substrate's material and the quality of the EB system, as well as the skill of the user, a resolution of EB lithography can be as small as 10 nm. In this work, EB lithography (ELS 3700-Elionix Co. Ltd.) was used to make squared windows on the SiO$_2$/Si substrate for forming perfect pyramidal tips on the Si substrate. Since the adhesive

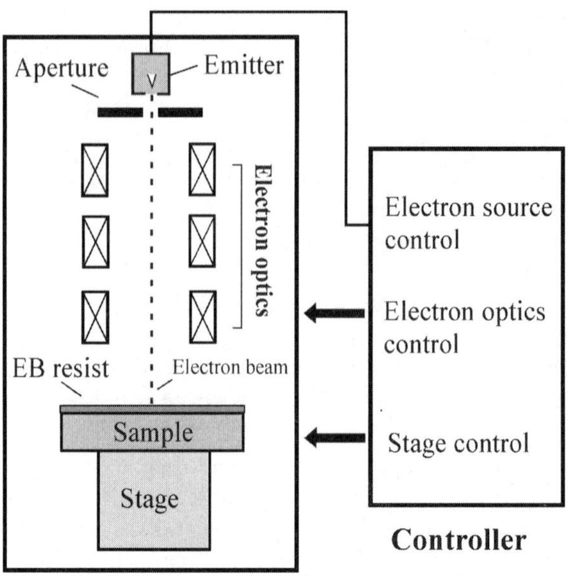

Figure 3.2 Simplified illustration of the electron beam lithography system.

between the EB resist and the substrate is not very strong, after EB lithography, the SiO_2 film was patterned by dry etching using fast atomic beam (FAB) etching. In the FAB system a high-energy, neutral atomic beam is utilized for etching. It provides perfectly vertical etching and eliminates charging effect during etching.

3.1.3   Focused ion beam milling

Focused ion beam (FIB) is a very effective tool for lithography, maskless implantation, nanoscale metal patterning and milling in integrated circuit (IC) repair. FIB utilizes a high-energy Ga ion beam to etch the desired parts with the help of scanning electron microscopy (SEM). FIB can also be utilized for local deposition of metal such as tungsten (W) with a $W(CO)_6$ gas source. In this work FIB (SMI-8100, Seiko Co. Ltd.) was used for making a cross-section of the microfabricated NSOM tip. FIB etching is often used to form aperture at the apex of the tapered optical fiber tip. On the first attempt at fabrication of an NSOM probe, FIB was used directly to form aperture at the apex of the tip.

3.1.4   Lift-off process

The lift-off process is very useful in making, for example, metal wires on a substrate. In some cases, metal patterns on the substrate can be formed by metal etching using photoresist patterns as masks. However, lift-off is more

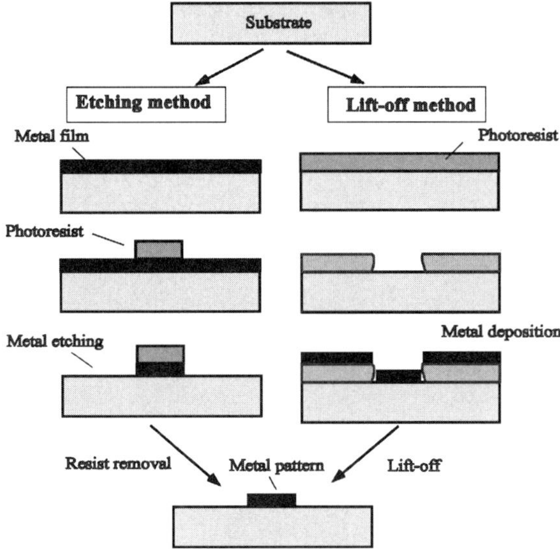

Figure 3.3 Principle of forming metal wire on a substrate by the etching and lift-off methods.

widely used. The principle of the lift-off and metal etching methods for forming metallic patterns on the substrate is shown in Figure 3.3. Normally, positive photoresist is used for the lift-off method. A solvent is used to dissolve the remaining positive photoresist underneath of the metal, and thus sacrificial layer is normally removed in acetone solvent and an ultrasonic vibration.

3.2 Thermal oxidation of silicon

Thermal oxidation is a very important technique in microelectronics and micromachining technology. Thermal SiO_2 is used as a perfect insulator, as a mask for etching Si or for diffusion of phosphor (P) or boron (B) into the Si, or as a sacrificial material for some surface micromachining technology. Thermal SiO_2 is also important for other applications, for example, optoelectronics and near-field probes. The most critical part of the NSOM probe is the aperture tip, which is made of thermal SiO_2.

SiO_2 can be grown on the Si surface in a stream of dry oxygen (dry oxidation) or steam of oxygen and water (wet oxidation) at an elevated temperature (600–1250°C). The high temperature aids the diffusion of oxidant through the surface oxide layer to the Si/SiO_2 interface to form the oxide by the following reaction:

$$Si + O_2 = SiO_2 \text{ (dry)} \tag{3.1}$$

$$Si + 2H_2O = SiO_2 + 2H_2 \text{ (wet)} \tag{3.2}$$

When an oxide layer of thickness of d_{ox} is grown on the Si surface, an Si thickness (d_s) of 0.46 d_{ox} was eaten.

$$d_s = 0.46 \, d_{ox} \tag{3.3}$$

Due to the molecular volume mismatch and thermal expansion differences, a compressive stress is generated at the interface Si/SiO_2. The dynamic of the thermal oxidation on planar Si is explained well by the Deal-Grove theory.[6] The relationship between the thickness of the oxide layer (d_{ox}) and the oxidation time (t) according to Deal-Grove model is

$$d_{ox}(t) = (A/2)\{[1 + 4(t + \tau)B/A^2]^{1/2} - 1\} \tag{3.4}$$

$$A = 2D(1/k_s + 1/h) \tag{3.5}$$

$$B = 2DC/N; \, \tau = (x_i^2 + Ax_i)/B \tag{3.6}$$

Where k_s is Si oxidation rate constant, which is a function of temperature T, oxidant, crystal orientation and doping of the substrate; D is oxidant diffusivity, that is a function of temperature and oxidant; h is gas phase mass transport coefficient in the solid in terms of concentration; C is equilibrium oxidant concentration in the oxide; N is the number of molecules of oxidant per unit volume of oxide (2.2×10^{22} cm^{-3} for dry oxygen); x_i is initial oxide thickness. From Equations 3.4, 3.5 and 3.6, for a thin oxide when the oxidation time ($t << A^2/4B$), the oxidation thickness is expressed approximately as a linear function of oxidation time t:

$$d_{ox}(t) \cong (B/A) \, (t + \tau) = C/N(1/k_s + 1/h)^{-1}(t + \tau) \tag{3.7}$$

where (B/A) is the linear rate constant ($\mu m/h$). For the thick oxides (long oxidation times with $t >> \tau$ and also $t >> A^2/4B$, the oxidation thickness is expressed approximately as a parabolic function of the oxidation time:

$$d_{ox}^2 \, (t) = B \, (t + \tau) = 2D(C/N) \, (t + \tau) \tag{3.8}$$

where B is the parabolic rate constant. The oxidation growth rate is dependent on the orientation of the Si crystal. This effect is involved in the linear oxidation rate constant (B/A) from Equation 3.7. The linear oxidation rate constant depends on the orientation of the Si substrate. For example, the linear oxidation rate constant of a Si (111) plane is 1.7 times larger than that of a (100) plane, i.e, the (100) surface oxidized 1.7 times more slowly than the (111) one. This can be explained as the difference in the bond density at the surfaces, for example, for (111) and (100) surfaces, the bond density of 1.76×10^{14} cm^{-2} and 6.77×10^{14} cm^{-2}, respectively, was found.

According to Dean-Grove, the oxidation process follows the following stages:

- Species of oxidant are transported from the bulk of oxidizing gas to the outer surface of the substrate
- Species are transported by diffusion across SiO_2 film to the SiO_2/Si interface
- Species reacts at the Si interface to form a new SiO_2 layer according to Equations (3.1 and 3.2)

With these processes, SiO_2 is gradually grown on the Si substrate. The Dean-Grove theory clearly explains thermal oxidation on planar Si. However, for a nonplanar structure of Si substrate, such as at the convex or concave corners, an anomaly of the oxidation is found. The anomaly created by thermal oxidation on nonplanar structures is a key point of the technique for forming an aperture and aperture array for NSOM applications.

A simplified illustration of the conventional thermal oxidation system is shown in Figure 3.4. Oxidation temperature with approximately 1°C error can be achieved. For dry oxidation, the purified dry oxygen flows directly through the oxidation furnace (V1, V2 valves: open; V3, V4 valves: closed). For wet oxidation, the purified dry oxygen is steamed with the vapor of boiled water before flowing through the oxidation furnace (V2, V3, V4 valves: open; V1 valve: closed).

3.3   Metallization

Two basic metallization methods were used in this work, electron beam evaporation and RF-magnetron sputtering.

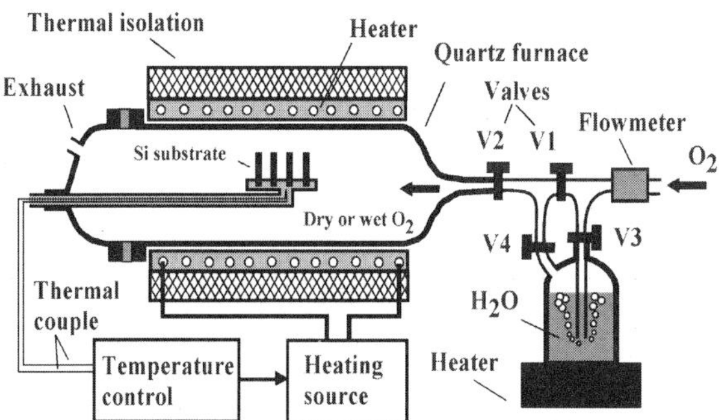

Figure 3.4 Simplified illustration of the Si thermal oxidation: Dry oxidation – V1, V2 valves open; V3, V4 valves closed. Wet oxidation – V2, V3, V4 valves open; V1 valves closed.

3.3.1   Sputtering technique (RF sputtering)

The sputtering technique was chosen for metallization due to several advantages: 1. wider choice of materials to work with; 2. better step coverage; and 3. better adhesion to the substrate. Normally, the chamber is evacuated below 10^{-6} torr and purified Argon (Ar) gas is then injected into the chamber to a pressure of 10^{-3} torr. During sputtering, the metal target, at a high negative potential, is bombarded with positive Ar ions, created in a plasma. The target material atoms are knocked out of the target and deposited on the substrate placed on the anode. The thickness of the metal film is controlled by optimizing: the power of the RF plasma; sputtering time; Ar gas pressure; target–sample distance. The sputtering technique with Al, Cr, Au, Al, Pt, Ti, W and Mo targets was used in this work. Sputtering was also used for deposition of insulators such as Si_3N_4 or SiO_2 film on to the Si substrate.

3.3.2   Electron beam evaporation

Electron beam evaporation was used for making electrodes and for the lift-off process. The technique is based on the conventional thermal evaporation method where the desired metal is melted by bombardment of a high intensity electron beam emitted from a thermionic filament in a high vacuum (10^{-6}–10^{-7} torr). The electron beam is magnetically directed onto the metal source. The metal source is melted and evaporated onto the sample. A quartz vibration sensor is installed into the chamber permitting the monitoring of the thickness of the metal film. In this work, the technique for forming Pt, Ti, Cr, Au, Ni film was employed.

3.4   Silicon etching

In this work (100)-oriented Si substrates and wet Si etching techniques (Si bulk micromachining) were employed. It is known that the etching of the Si (100) is highly anisotropic and with several anisotropic wet etchants, the etching rate at {111} plane is much slower than that of other planes ({100} or {110}). The anisotropic etching of Si (100) has been effectively used in forming pyramidal etch pits, thin membranes, microchannels, trenches, holes and mesa structures. A wide variety of etchants have been used for anisotropic etching of Si (100), such as alkali aqueous solution of KOH, NaOH, tetramethyl ammonium hydroxide (TMAH) and ethylene diamine pyrochatechol (EDP).

All experiments of Si (100) wet etching in this work were done in a TMAH $(CH_3)_4NOH$, 25% etching for several reasons 1. the selectivity of etching Si and SiO_2 mask is very high (around 5000:1 at 80°C); 2. the etched Si surface is flat; 3. the etching rate is quite stable (0.5 $\mu m/min$, for TMAH 25%, 80°C), which facilitates control of the thickness of the Si cantilever beam; 4. safety level is high and is a CMOS compatible process. The main characteristic of the anisotropic wet etching of the Si (100) substrate is sketched in Figure 3.5. The lower in etching rate at the {111} plane compared to that of

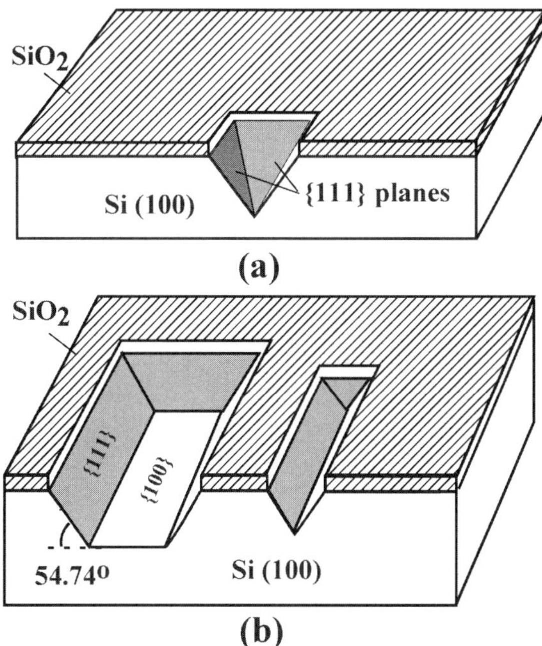

Figure 3.5 Anisotropic wet etching of (100) oriented Si wafer.

other planes leads to pyramidal, V or U groves with a base angle of 54.74°
as shown in Figure 3.5. If the SiO$_2$ opening is perfectly square or round, the
etching will stop at the intersection of four {111} planes that form a perfect
pyramidal shape with the depth about 0.7 times compared to the mask
opening size (Figure 3.5a). If the opening mask is not square, a channel
limited by four {111} planes and a rectangle {100} plane at the top can be
formed (see Figure 3.5b). If the etching goes on, it will eventually stop when
the top rectangle plane becomes a line at the top. The desired structures can
be achieved by optimizing the designed mask and the etching rate on the
orientation of the wafer.

An experimental setup of the TMAH wet etching of silicon in this work
is illustrated in Figure 3.6. The etching temperature is controlled using a Pt
temperature sensor, a temperature controller, and a heater. To increase the
uniformity of the temperature inside the etching chamber and to facilitate
the dissolution of the Si etching, a magnetic stirring system is used.

The mechanism of the anisotropic wet etching of Si has been explained
with different models and reported in several references.[7-9] The process of
etching Si is understood as the following.[9] Si atoms at the surface react with
hydroxyl ions and generate 4 electrons as:

$$Si + 2OH^- \rightarrow Si(OH)_2^{++} + 4e^-$$

(3.9)

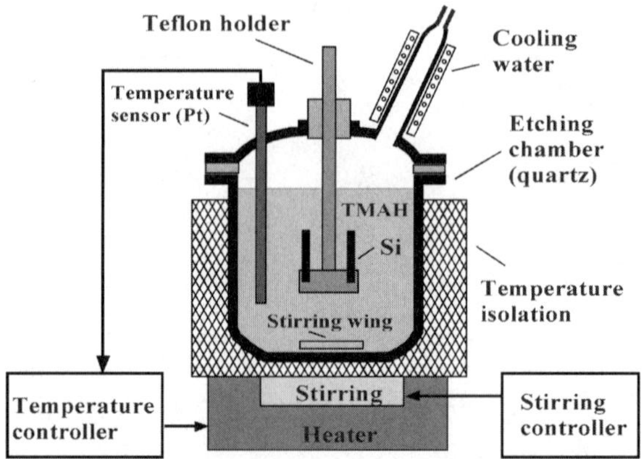

Figure 3.6 Experimental setup of the TMAH wet etching of silicon.

Electrons in (3.9) leave the solid, return to the etchant and react with water as:

$$4H_2O + 4e^- \rightarrow 4OH^- + 2H_2 \qquad (3.10)$$

The sequence (3.9, 3.10) is repeated and Si is etched. A heavy doping boron layer ($P^{++} > 7 \times 10^{19}$) is an etching stop layer with wet etching because the electrons in the sequence will recombine with holes in the P^{++} layer and break the sequence between the solid and the liquid. The space charge region in a p-n junction can also be used to stop the wet etching of silicon. The mechanisms of anisotropic etching and etching stops are explained with several models. For instance, Kendall[7] ascribed the extreme differences between the very low etching {111} planes and the much faster {110} and {100} planes by the fact that the {111} planes oxidized much faster than other planes in the etchant. The rapid oxidation of the {111} planes prevented the surface from etching, but the other planes did not oxidize fast enough to protect the surfaces from dissolution. In this work only techniques of Si etching for fabrication of the NSOM probes are described.

3.5   Silicon oxide etching

SiO_2 etching is an important step in microfabrication and extremely important in this work. Commonly, SiO_2 wet etching is treated with solutions containing hydrofluoric acid (HF). The SiO_2 etching rate can be varied widely by adding different additives, such as nitric acid (HNO_3), orthophosphoric acid (H_3PO_4), or ammonium fluoride (NH_4F).[9] The dissolution rate of the SiO_2 depends on the etching temperature, pH of the solution, and the properties of the SiO_2, such as density, doping and stress. The mechanism for dissolving SiO_2 in solution is complicated and full understanding of the process has not yet been achieved.

In this work SiO_2 etching was done by dry (using the FAB etching as previously mentioned) or wet etching with buffer-HF (BHF) of 50% HF:40% NH_4F = 9:100. BHF wet etching is desirable because it etches using a moderate etching rate (about 100nm/min at room temperature) and markedly less undercutting (lateral etching) of the oxide under the photoresist layer. The etching mechanism using BHF solution has been reported by several different investigators.[10-15] It was confirmed that in a system of NH_4F/HF-containing solutions, there were several species, namely H^+, F^-, HF_2^-, NH_4^+, HF. These species were formed through the following relations:

$$HF \longleftrightarrow H^+ + F^- \tag{3.11}$$

$$NH_4F \longleftrightarrow NH_4^+ + F^- \tag{3.12}$$

$$HF + F^- \longleftrightarrow HF_2^- \tag{3.13}$$

The overall chemical reaction of the SiO_2 in the HF solution is normally understood as:

$$SiO_2 + 6HF \rightarrow 2H^+ + SiF_6^{-2} + 2H_2O \tag{3.14}$$

It was shown that depending on the pH of the etchant, for low concentration of NH_4F, the reaction is dominated by the HF_2^- species and by the HF for the higher concentration of NH_4F. This should be related to their differences in activation energies. The domination of the etching SiO_2 with HF when increasing the NH_4F is explained as the role of NH_4^+ in coordination with HF_2^-.[10]

3.6 *Anodic bonding and packaging*

Anodic bonding is one of the most important bonding techniques in microfabrication. Normally, anodic bonding is indicated for the bonding between the Si and glass with high alkali metal content (e.g., Pyrex glass). Anodic bonding is utilized at approximately 400°C under a small force and a high electric field with negative electrode potential on the glass side.

A simplified illustration of the anodic bonding system is shown in Figure 3.7a. The mechanism of the anodic bonding is as follows (see Figure 3.7b). Under a high electric field, positive ions (Na^+) in the glass substrate move toward the cathode, creating a space charge region at the interface Si/glass. This space charge region creates a strong electrostatic attraction between Si and glass that hold both pieces in place. Under this electric field, oxygen in the glass moves toward the interface where it reacts with Si to form a thin oxide layer at the interface. The bonding works at a high temperature to increase the mobility of the alkali ions. For some specific devices, the bonding

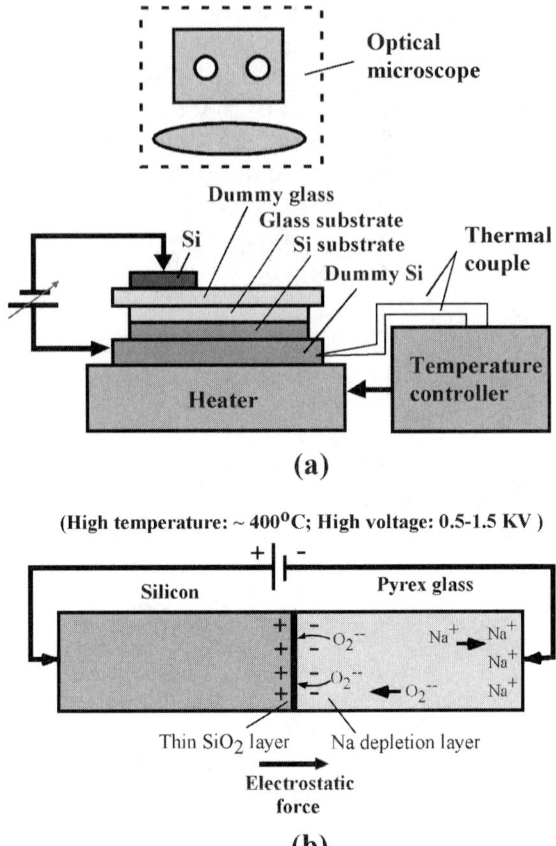

Figure 3.7 (a) Anodic bonding of silicon and Pyrex glass; (b) mechanism of the anodic bonding.

process is done in a vacuum with computerized control, which allows monitoring of the current of Na ions in the glass base.

Besides the abovementioned techniques, several other techniques in microfabrication (not presented here) were used in this work, such as boron doping to form P^{++} elements, ICP-RIE dry etching of silicon, plasma CVD for deposition of SiO_2, metal etching, YAG-laser machining and device dicing.

References

1. Madou, M., *Fundamentals of Microfabrication*, CRC Press, Boca Raton, FL, 1997.
2. Esashi, M. et al., *Micromachining and Micromechantronics*, Tokyo, 1992 (in Japanese).
3. Peterson, K.E., Silicon as a mechanical material, *Proc. of the IEEE*, 70, 420, 1982.

4. Kovacs, G.T.A., *Micromachined Transducers: Sourcebook*, McGraw-Hill, New York, 1998.
5. Rai-Choudhury, P., *Handbook of Microlithography, Micromachining, and Microfabrication, Volume 2: Micromachining and Microfabrication*, SPIE Press, Bellingham, WA, 1997.
6. Deal, B.E. and Grove, A.S., General relationship for the thermal oxidation of silicon, *J. Appl. Phys.*, 36, 3770, 1965.
7. Kendall, D.L., On etching very narrow grooves in silicon, *Appl. Phys. Lett.*, 26, 195, 1975.
8. Seidel, H. et al., Anisotropic etching of crystalline silicon in alkaline solutions, *J. Electrochem. Soc.*, 137, 3612, 1990.
9. Kovacs, G.T.A., Maluf, N.I., and Peterson, K.E., Bulk micromachining of silicon, *Proceedings of the IEEE*, 86, 1536, 1998.
10. Proksche, H., Nagorsen, G., and Ross, D., The influence of NH_4F on the etch rates of undoped SiO_2 in buffered oxide etch, *J. Electrochem. Soc.*, 139, 521, 1992.
11. Judge, J.S., A study of the dissolution of SiO_2 in acidic fluoride solutions, *J. Electrochem. Soc.*, 118, 1772, 1971.
12. Nielsen, H. and Hackleman, D., Some illumination on the mechanism of SiO_2 in HF solutions, *J. Electrochem. Soc.*, 130, 708, 1983.
13. Kikuyama, H. et al., Etching rate and mechanisms of doped oxide in buffered hydrogen fluoride solution, *J. Electrochem. Soc.*, 139, 2239, 1992.
14. Parisi, G.I., Tapered windows in SiO_2: effect of NH_4F:HF dilution and etching temperature, etch rates of doped oxides in solutions of buffered HF, *J. Electrochem. Soc.*, 124, 917, 1977.
15. Tenney, A.S. and Ghezzo, M., *J. Electrochem. Soc.*, 120, 1091, 1973.

part 2

Experimental results and discussion

chapter four

Fabrication of silicon microprobes for optical near-field applications

4.1 Overview of micromachined optical near-field probes

In order to overcome the drawbacks of the optical fiber based probes as presented in Chapter 2, several approaches for fabrication of the NSOM probes using Si micromachining technology have been proposed. This section will give a short overview of some of them. All updated contributions in the field, however, could not be covered. The main contributions of micromachined optical near-field probes are briefly discussed. Every technique has its advantages and disadvantages. The comparisons presented are intended for readers in the hope that they can find useful information for their specific purposes.

The first attempt to fabricate an NSOM probe was accomplished by van Hulst et al.[1-3] A microfabricated Si_3N_4 cantilever with a transparent pyramidal Si_3N_4 tip was used in a photon scanning tunneling microscopy (PSTM) configuration. The evanescent field was generated at the surface of the sample by the total internal reflection. The Si_3N_4 tip converted the evanescent wave into a propagation wave. A long-distance working lens (40X, 0.5 NA) connected with a photomultiplier tube (PMT) was used to collect the diffracted light. A laser diode and a quadrant detector were used to monitor the bending of the cantilever for control of the tip–sample distance (AFM technique). Using a laser beam of far-field light, with power of 5 mW, a 0.1 nW, near-field intensity was achieved in the far-field when the tip was almost touching the sample. Using a very sharp Si_3N_4 tip, a resolution of 40 nm was reported.[1] This is a very simple and effective method because the tip–sample distance can be controlled by the AFM technique and the commercially available Si_3N_4-AFM cantilever can be used. The topography and optical near-field imaging can be recorded simultaneously. However, no aperture was used in this work and, as mentioned in Chapter 2, it is likely to be difficult to eliminate the huge background signal on the surface of the sample.

Oesterschulze et al. proposed a technological solution for manufacturing a hollow metal tip with an aperture integrated on an Si cantilever.[4-7] After forming a pyramidal etch pit on an Si (100) wafer by anisotropic etching in a KOH solution, the wafer is thinned by etching in the KOH until forming an aperture at the apex of the etch pit. A 120 nm thick Cr layer is deposited on the upper side of the etch pit to form the desired hollow pyramidal tip. Finally, a reactive ion etching was used for further thinning of the Si, and for releasing the Si cantilever with a hollow Cr tip that had the aperture at its apex. The cantilever with about 80–120 nm Cr-tip aperture was successfully fabricated and utilized as an AFM/NSOM probe for topographic and subwavelength optical imaging of several surfaces.[7] With this process the size of the metallic aperture is determined by the size of the aperture formed at the Si molding and the thickness of the Cr film. This is a batch fabrication process that permits producing many cantilevers and apertured tips in one wafer. Optical throughput is improved in comparison to the conventional optical fiber probe. However, the technique faces difficulty in controlling the aperture size. The inhomogenneity in thickness of the Si substrate and that of the etched Si surface, as well as the discrepancy in size of the etch pits, may affect the reproducibility in the size of the apertures.

Quate et al. developed a technique for forming a microfabricated probe with a small aperture for ion-conductance microscope.[8] The probe tip is made of a hollow, heavily boron doped Si tip. The fabrication process can be briefly explained as follows. First an Si protrusion is formed on the (100) oriented-Si wafer. The wafer is then doped with boron for the desired thickness. By completely etching the underneath of the undoped Si parts, an aperture at the center of the protrusion is created. The hollow tips with typical aperture sizes of 250 nm were obtained by this technique. It is seen that the shape of the tip formed by this technique is quite reproducible, however, the reproducibility in the size of the aperture is somewhat difficult to control. Size of the aperture depends on the size of the Si protrusion at the apex and the thickness of the boron doped layer. Nevertheless, the technique could be advanced in the manner of mass production. The probe is suited for an ion-conductance microscope. For NSOM applications, a metallic opaque layer may be needed because of a slight transparence in the boron doped Si tip.

Another approach to fabrication of a metallic aperture at the apex of a transmissive Si_3N_4 tip is that of Drews et al.[9,10] The Si_3N_4 sharp tips are produced from a chemical vapor deposition–Si_3N_4 layer by isotropic dry etching in a CF_4–plasma. An Al film of approximately 300 nm thick is coated on to the tips. The aperture is formed by selectively etching the Al at the apex of the Si_3N_4 tip until the tip's apex becomes transparent. The etching process in this technique is somewhat the same as the selective etching technique of the optical fiber based NSOM probes. An optical throughput of 10^{-3}–10^{-4} for aperture of 100–nm diameter was reported. However, this technique also suffers from low reproducibility of the size and shape of the tip and the aperture.

By combining Si micromachining and electron beam nanolithography Zhou et al.[11] demonstrated a technique for making an aperture at the tip. An Si substrate with a pyramidal protrusion was used as a mold for deposition of a low stress silicon nitride layer (500 nm in thickness). The aperture at the center of the protrusion was defined by electron beam lithography and created by dry etching. After removing the Si mold, a 130–150 nm thick Al was coated by evaporation. A minute aperture with a size of 150 nm with 10^{-3} optical throughput was reported using the described technique.[11] This process is not simple and it is difficult to form a very small aperture at the desired location at the apex of the tip.

Another technique for manufacturing an Si microcantilever with a microscopic photodiode (PN-junction) integrated at its end has been developed by Akamine et al.[12] Optical near-field imaging with a resolution of 20 nm is reported. The distance between the photodiode and the sample is kept constant by the AFM technique. With this probe, only PSTM configuration can be utilized and a sharp tip is needed to improve the resolution. Moreover, the detected near-field signal by the detector may be affected by the signal of the laser for atomic force detection.

This study was originally intended to create an aperture at the apex of a Cr coated Si tip (heavily doped with boron) on an Si cantilever by focused ion beam milling (see Figure 4.1). An aperture with a diameter of smaller than 200 nm was fabricated with this milling technique.[13] However, this method was no longer feasible. There was difficulty in forming a very small aperture at the desired position of the tip with ion beam milling. With a high-performance FIB machine, available today, it would be possible to form aperture sizes of smaller than 100 nm. This technique, however, is time-consuming and expensive.

A modified fabrication process adapted from Oesterschulze's method is plotted in Figure 4.2.[13] First the Si cantilever's shape and pyramidal etch pit was formed by photolithography, SiO_2 patterning, and Si wet etching in a tetramethyl ammonium hydroxide (TMAH) etchant (steps a–c). Next, the Si wafer is etched from the lower size of the TMAH until forming a small aperture at the etch pit (steps d, e). A thin Cr film of 80–120 nm is then entirely deposited on the upper side of the sample (step f). Finally, the Si wafer is further thinned by etching in the TMAH until releasing the Si cantilever with an integrated hollow Cr tip (step g). Typical SEM images of the Si cantilever with 140×100 nm^2 aperture at the apex of the hollow Cr tip is shown in Figure 4.3. The aperture or aperture array formed by this technique is not very reproducible because of the inhomogeneity of the Si substrate, the etched Si surface, as well as the discrepancy in size of the etched pits.

Another technique based on the field evaporation effect was also utilized (see Figure 4.4). A SiO_2 tip integrated at the end of the Si cantilever was microfabricated (detail of the fabrication of SiO_2 tip on the Si cantilever will be explained in next section). A thin Cr film was entirely coated on the SiO_2 tip to form an opaque screen. The Cr coated tip is then brought into contact

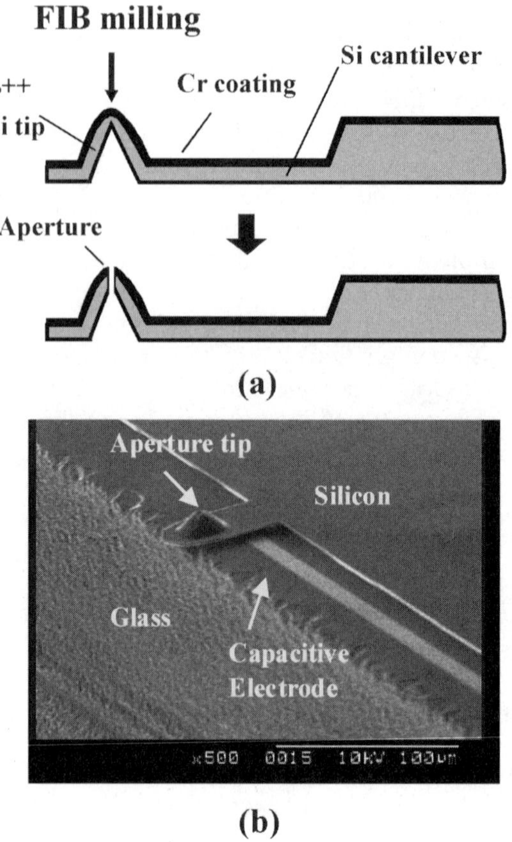

Figure 4.1 Microfabricated NSOM probe based on the Si cantilever with the Cr-coated, boron-doped Si tip. Aperture was formed by FIB milling, (a) schematic view; (b) SEM image of the fabricated probe.

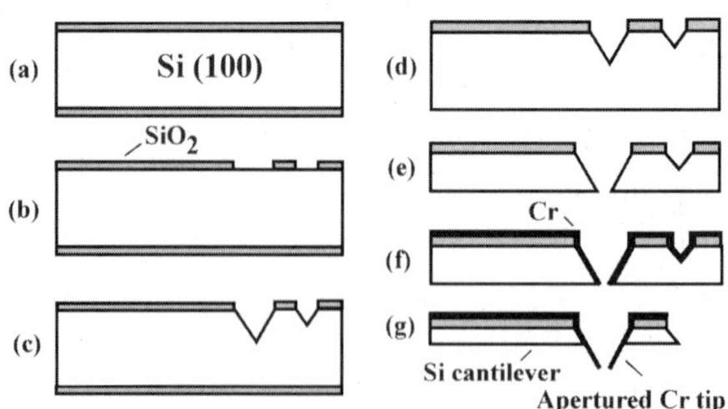

Figure 4.2 Fabrication process of forming a hollow metal tip integrated on an Si cantilever for NSOM probe.

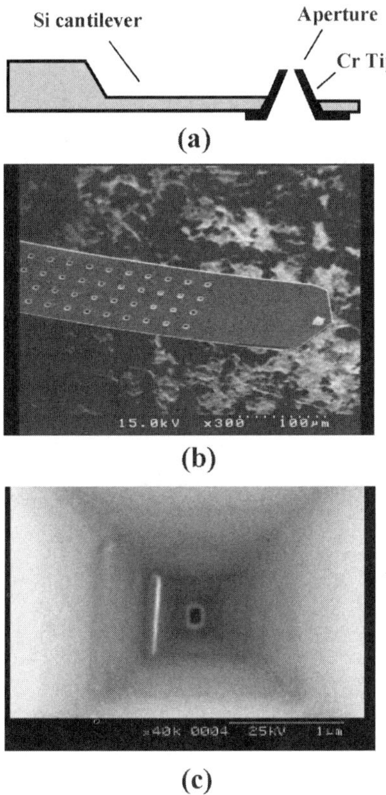

Figure 4.3 (a) Schematic view of the NSOM probe with a hollow Cr tip on the Si cantilever; (b) SEM images of the microfabricated NSOM cantilever; (c) close-up view of the aperture.

with the conductive surface (Cr coated glass substrate, in this case) with the help of the AFM system. By applying a suitable voltage between the cantilever and the conductive substrate, Cr atoms around the apex are field evaporated due to a high electric field leaving the aperture. Typical SEM images of the fabricated cantilever with about a 120 nm aperture that was formed by applying 10 pulse of 1V for a 100 ms pulse duration between the tip and Cr coated glass base, is shown in Figure 4.5. Aperture formed by this technique, however, is not acceptably reproducible. A number of parameters can affect the formation of the aperture, such as thickness of the opaque layer, contact area between the tip and the conductive substrate, and applied electric field. The technique also suffers from low production yield. It is difficult to control the size and shape of the aperture, but this technique may be useful in the fabrication of apertures at the apex of metal coated, commercialized SiN cantilevers.

Several other techniques for creating apertures were recently proposed, for example, the solid-solid diffusion,[14] or using a metal coated solid quartz

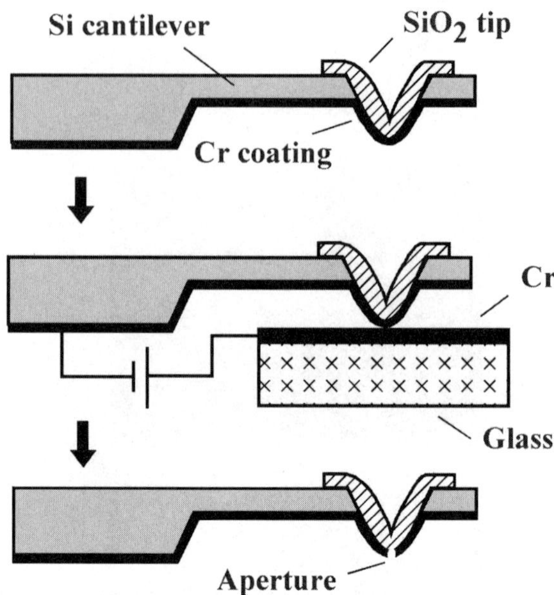

Figure 4.4 Schematic diagram of forming aperture using the field evaporation technique.

tip on the Si cantilever and reactive ion etching.[15,16] These methods can be advanced with improvements in reproduction. Recently, a method for fabrication of the aperture NSOM probes in a batch process was reported by Oesterschulze et al.[17] The technique is almost identical to the technique presented here. The aperture is formed at the apex of the SiO$_2$ tip by etching.

In summary, the most critical parts of the NSOM probe are the aperture at the apex of the tip and the tip itself. Many groups have been searching for different ways to fabricate a small aperture at the apex of the tip. A simple process to fabricate a small aperture by mass production with high reproducibility and a high-performance NSOM probe has not yet been achieved. This study presents a very simple technique called low temperature oxidation and selective etching (LOSE) to form a very small aperture (as small as 20 nm) at the apex of the thermally created SiO$_2$ tip in a batch process with high reproducibility. The method is very effective and applicable not only for optical near-field applications but also for other purposes. In this chapter details of the LOSE technique are presented.

4.2   Design of the probes

One of the advantages of Si micromachining is the flexibility in designing of the devices. In this work two types of probes are proposed, AFM/NSOM and capacitive-AFM/NSOM (as shown in Figures 4.6 and 4.7, respectively).

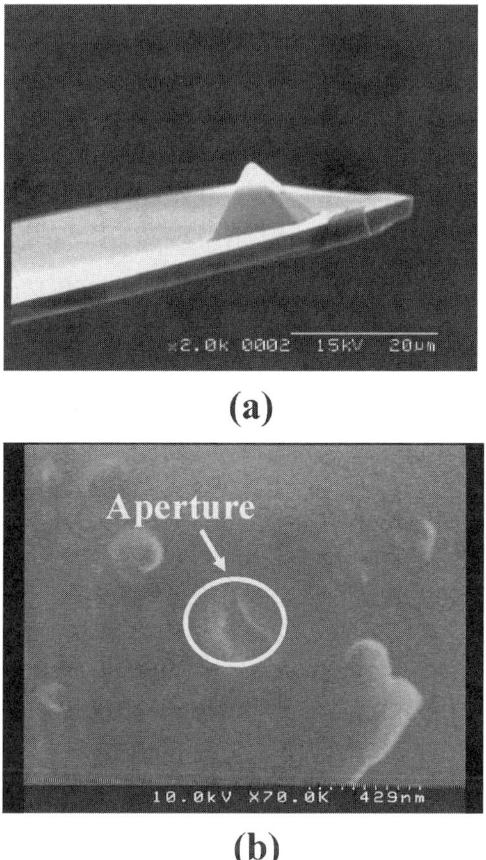

Figure 4.5 (a) SEM images of the microfabricated NSOM cantilever and (b) close-up view of the aperture formed by field evaporation technique.

Sixteen Si cantilevers on an Si piece of 2×2 cm^2 with different sizes of 100–900 μm length, 60–200 μm width, 3–5 μm thick corresponding to spring constants of 0.26–175 N/m and resonant frequencies of 5-625 KHz were designed. The current design is specialized for 2×2 cm^2 wafer process; however, it can be extended for larger wafer process.

A transparent thermal SiO$_2$ tip with a nanoscale-sized aperture at the center is formed at the end of the Si cantilever. The thermal SiO$_2$ tip was chosen because of the relationship to the critical technique of forming a small aperture. Moreover, the thermal SiO$_2$ tip shows good optical transparence in a wide range of wavelengths and strong mechanical wear which is suited for optical near-field applications.

The spring constants (k) and resonant frequencies (f$_o$) of cantilevers with different sizes are calculated according to the well-known equations following.

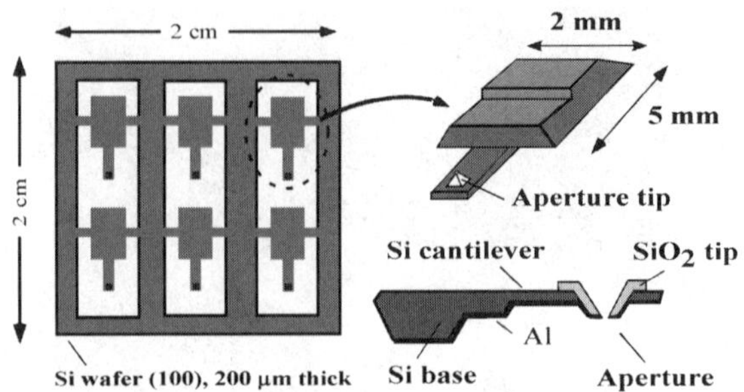

Figure 4.6 Design of the AFM/NSOM probe.

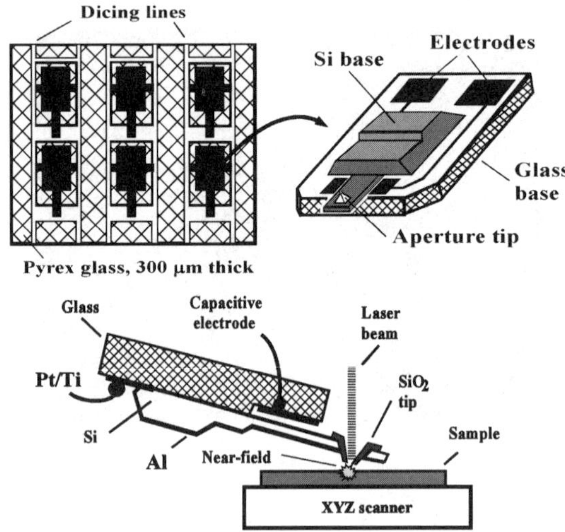

Figure 4.7 Design of the capacitive-AFM/NSOM probe.

$$f_o = (1/2\pi)\sqrt{(k/m)} \quad (Hz) \tag{4.1}$$

$$k = Ewt^3/(4L^3) \quad (N/m) \tag{4.2}$$

$$m = \rho wtL \quad (Kg) \tag{4.3}$$

where E is Young's modulus of crystal Si, $E = 1.4 \times 10^{11}$ N/m²; L, w, t are the length, width and thickness of the cantilever, respectively; m is mass of the cantilever; ρ is density of Si, $\rho = 2330$ Kg/m³. The calculated resonant frequencies and spring constants of the cantilever with different sizes are shown in Figures 4.8 and 4.9, respectively. Since the designed probe is based on the Si AFM cantilever, the tip–sample distance can be kept at a constant value by utilizing the AFM techniques in contact or dynamic modes with any AFM system.

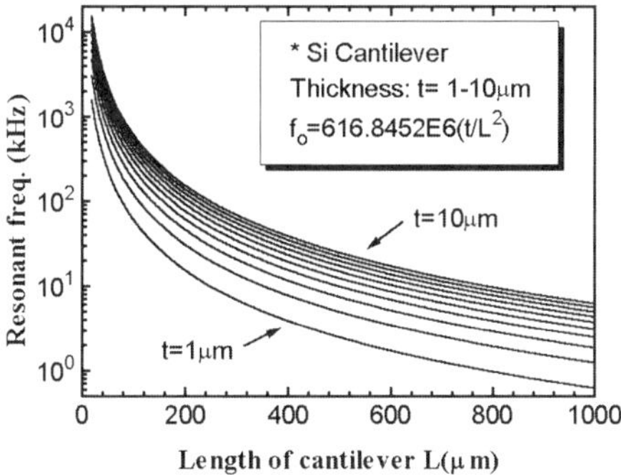

Figure 4.8 Resonant frequencies (f_o) of the Si cantilever with different sizes as a function of the cantilever's length (L).

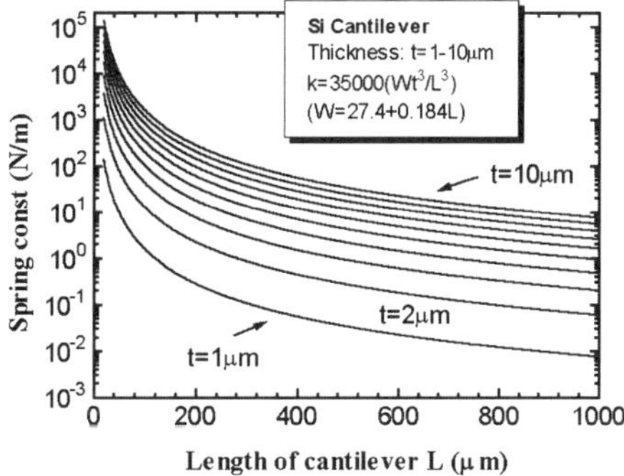

Figure 4.9 Spring constant (k) of the Si cantilever with different sizes as a function of the cantilever's length (L).

In dynamic mode two detection methods are usually used, the slope and the frequency modulation. In the slope detection method, the cantilever is mechanically vibrated near the resonance; the force between the tip and sample is detected from the change of the vibration amplitude. In the latter case, the cantilever vibrated at its resonance; the force is detected from the change of the resonant frequency. The slope detection method was employed for this probe. The minimum detectable force is limited by thermal noise that is proportional to $\sqrt{k}/f_o$. Therefore, low-noise AFM can be achieved by choosing a cantilever with a low spring constant and a high resonant frequency. Normally, the frequency of >10 kHz is required in AFM.

In contact mode the tip is held in contact with surface by a tracking force small enough to avoid surface damage (10^{-9}–10^{-7} N). A spring constant in the range of 10^{-2}–10^2 N/m is desired in contact mode. In dynamic mode the Q factor of resonance should be high because the cantilever will be vibrated near its resonance and a force gradient will cause the change in amplitude (the force gradient could also cause the change of spring constant as $k = k_0$ + dF/dz, so that it affects the resonant frequency and changes the amplitude). The higher the Q factor, the higher the sensitivity in noncontact mode. However, in contact mode, Q factor is not important.[18] With the design shown in Figure 4.6, the probe serves as a conventional AFM and NSOM both in contact or noncontact modes.

Due to the advantages of the Si micromachining technique, a novel probe of capacitive-AFM/NSOM is also feasible (see Figure 4.7). A narrow gap of 1–3 µm is formed between the Si cantilever and Pyrex glass base. A capacitive electrode is formed on the glass base. The Si wafer and glass substrate is anodically bonded. This capacitive–AFM/NSOM structure provides the means for sensing deflections of the cantilever and supplies the force for electrostatic actuation.[19] Normally the defection of the cantilever caused by the atomic force between the tip and sample is detected by optical or piezo-electric deflection.[20] With the capacitive–AFM/NSOM, several advantages can be addressed as the following:

- Deflection of cantilever can be detected electrically so that it can eliminate the complicated optical arrangement.
- It is possible to control the tip–sample distance capacitively and to actuate the cantilever vibration by applying an external voltage.
- In dynamic mode it is possible to keep force-balance by electrostatic force and avoid the jumping effect caused by the attractive force between the tip and the sample.

The most critical part of the probe is the ultra-small aperture at the apex of the tip.

4.3 *Principles of the fabrication process*

The technique of forming small aperture at the apex of the thermally produced SiO_2 tip is based on the anomaly of the oxidation grown on nonplanar Si at low temperature. As described in Section 3.2, the thermal oxidation in the planar Si structure was explained quite well with the Deal-Grove model using two oxidation regions. In the first (linear) region, oxidation is the interfacial reaction of the oxygen-containing species with the Si surface. Beyond a critical thickness of the oxide layer, the second region where the diffusion of species across the already grown oxide becomes dominant and the reaction follows parabolic rate law.

For nonplanar Si (convex or concave structures), the oxidation is quite different from the Deal-Grove theory. Since 1982,[21] Marcus proved that the oxide at the convex and concave corners of a nonplanar Si becomes thinner than other parts (as shown in Figure 4.10a) especially at low temperature oxidation (below 1050°C). The reduction of the oxidation rate at a low temperature at the concave corner has been explained by the geometry of the structure. A certain flux of oxidation species moves to the concave corner and

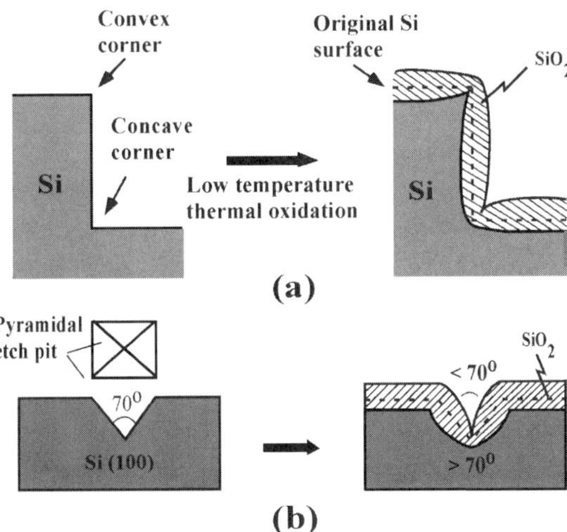

Figure 4.10 (a) Anomaly of the thermal oxidation at the convex and concave corners of the Si substrate; (b) anomaly of the low temperature grown SiO_2 at the etch pit of the (100)-oriented Si substrate.

oxidizes a larger interface compared with an oxide region bounded by the flat surface. Accordingly, the concentration of the oxidant moving to this corner must be lower, i.e., oxide layer at the concave corner becomes thinner. This explanation seems reasonable for the case of the concave corner. However, it cannot explain the reduction of the oxidation rate at the convex corner, which, at a low temperature, has been attributed to higher compressive stress induced there compared to other regions. The compressive stress is predicted to cause a depressing effect on the diffusivity of the oxidant into the interface Si/SiO_2 and slow down the oxidation rate. The mentioned effect at the convex and concave corners becomes less effective at higher oxidation temperatures. This was explained as the relaxation of stress due to the viscous flow of SiO_2 at higher temperatures.

The reduction of the oxidation rate also happens at the etch pit formed on the Si (100) substrate as schematically shown in Figure 4.10b. Before oxidation, the etch pit has a perfectly pyramidal shape that is the intersection of four {111} planes. As drawn in Figure 4.10b, a cross-section of the etch pit has a top angle of 70°. After oxidation at a low temperature at a certain thickness, the upper contour of the oxide layer becomes a convex bow shape with the top angle much smaller than 70°, whereas the lower contour becomes a concave bow shape with the top angle much larger than 70° in the cross-sectional view. Consequently the thickness of the oxide at the top of the pyramidal etch pit is much thinner than other parts. The shape and the difference in thickness of the SiO_2 at the top of the pyramidal etch pit and other parts strongly depends on the oxidation temperature and oxidation time. It was reported that the mentioned effect stood out in relief at the oxidation temperature of 950°C.[21]

The main reason for the mentioned effect both at the convex and concave corners is predicted to be due to the contribution of the compressive stress. The influence of the stress and the effect of relaxing stress at higher temperature oxidation have been calculated,[21-26] supporting the hypothesis of stress.

The effect of reduction of oxidation rate at the convex corner at low temperature, even though is not yet clear, was effectively utilized to make a sharp tip for AFM or field emitter application.[27] The principle of the sharpening Si tip is schematically shown in Figure 4.11a. After oxidation at a low temperature and the etching of the grown SiO_2 film, the protrusion on Si substrate becomes much sharper. The sharpness of the tip depends on the initial shape of the protrusion and the thickness of the SiO_2 layer. In some cases, the process shown in Figure 4.11a is repeated to get the desired sharpness.

The effect of lowering the oxidation rate at the concave corner has been utilized for making a mold for deposition of Si_3N_4 film with a sharp tip for AFM applications.[28,29] As schematically shown in Figure 4.11b, the Si with pyramidal etch pit is oxidized at a low temperature for a certain thickness.

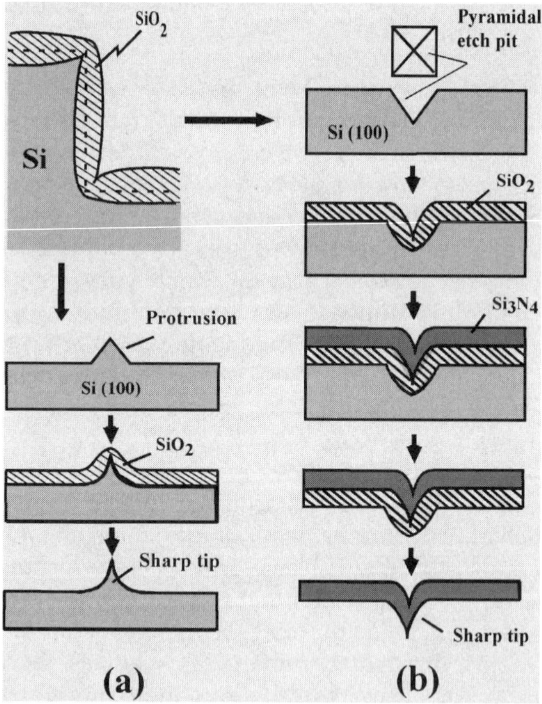

Figure 4.11 (a) Application of the anomaly of the low temperature grown SiO_2 at the convex corner for making a sharp Si tip; (b) application of the anomaly of the low temperature grown SiO_2 at the concave corner to form a mold for making a sharp Si_3N_4 tip.

Next, an Si_3N_4 film is deposited on the oxidized Si mold. The Si_3N_4 cantilever with a sharp tip is released by removing the mold. These effects are now effectively utilized and have become a routine technique for sharpening tips.

In another approach, we applied the anomaly of the oxidation at the pyramidal etch pit for making an ultra-small aperture or aperture array with the so-called low temperature oxidation and selective etching (LOSE) technique. Details of the fabrication technique has been reported in references 30 and 31 and will be explained in the next section.

4.4   Details of the fabrication process

The fabrication technique is based on the fundamental processes of Si micromachining: pholithography, SiO_2 patterning, Si etching, metalization, metal etching, anodic bonding, dicing and importantly, the anomaly of the oxidation growth at 950°C at the pyramidal etch pit. The fabrication process of the probes using the LOSE technique is schematically shown in Figures 4.12 and 4.13. For the AFM/NSOM probes, the Si process shown in Figure 4.12 is used. For the capacitive-AFM/NSOM probes, the whole process including the Si process, glass process and bonding process in Figures 4.12 and 4.13 are utilized.

4.4.1   Si process (see Figure 4.12)

An Si wafer of (100) orientation, 200-μm-thick, n-type, resistivity of 0.01–0.1 Ωcm was chosen as a starting material. The wafer is cut into 2×2 cm² pieces. First, the Si wafer is cleaned and thermally oxidized. Next, photolithography

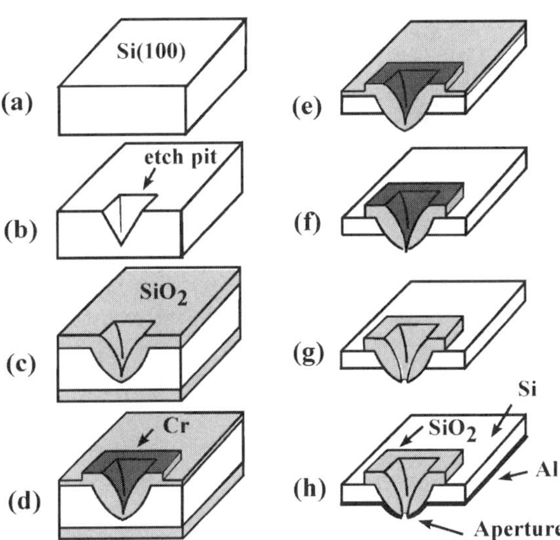

Figure 4.12 Si fabrication process of the AFM/NSOM probe using the LOSE technique.

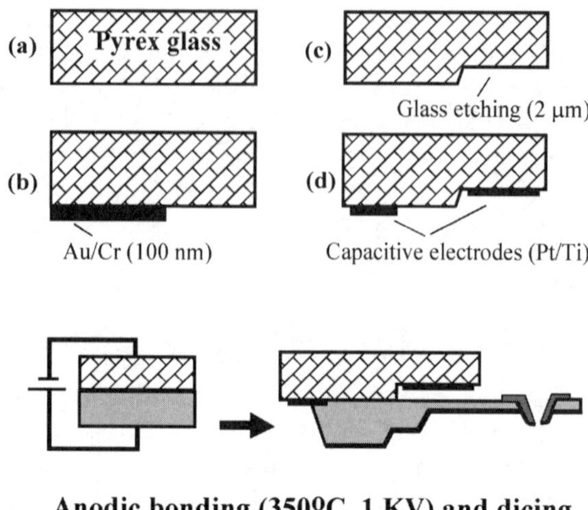

Anodic bonding (350°C, 1 KV) and dicing

Figure 4.13 Fabrication using glass and bonding processes of the capacitive-AFM/NSOM probe.

is done to define the mask patterns for etching SiO_2. Next, the pyramidal etch pit and structure of cantilevers are formed by anisotropic etching of Si in a tetramethyl ammonium hydroxide (TMAH) solution at 80°C (steps a–b). Next, an SiO_2 film is thermally grown by wet oxidation at 950°C with a thickness of around 1 μm (step c). Subsequently, a 100-nm-thick chromium (Cr) film is deposited on the upper side of the oxidized pit as a protective pattern (step d). This Cr pattern can be formed either by sputtering and etching or the lift-off technique. Cr was chosen because of its good adhesion with the SiO_2. Using this Cr pattern as a mask, the SiO_2 of about 0.6 μm in thickness is partly etched as shown in step d. The remaining oxide of about 0.4 μm in thickness is needed for the further processing steps. Next, the Si wafer is anisotropically etched from the lower side in the TMAH solution until forming Si cantilever with SiO_2 tip at the end of the cantilever (step e). The wafer is then dipped into the buffered-HF (50%HF: 40%NH_4F, 9cc:100cc) solution for selective etching until the Cr protrusion comes into sight (step f). At this step only the exterior wall of the SiO_2 tip is etched in the BHF while the interior wall is protected by the Cr pattern. Next the protective Cr layer on the upper side of the tip is etched out in a conventional Cr etchant (step g) to create the aperture. Finally, an Al/Cr or Al film with the thickness of approximately 80–120 nm is deposited onto the lower side of the cantilever to form an opaque layer (step h).

It should be emphasized that two points are important in the LOSE technique. First, the SiO_2 tip should be formed by thermal oxidation in wet oxygen at 950°C (step c) so that the thickness of the SiO_2 film at the apex of the tip will be thinnest compared with the other parts due to a large compressive stress. It has been experimentally observed that after thermal oxidation at 950°C for about 1 μm in thickness, the SiO_2 film at the apex of the tip was approximately 25–35% compared with that of the flat region. Second, the aperture at the apex of the SiO_2 tip should be formed by etching the SiO_2 tip in the BHF or dry etching only from the lower side using a protective layer on the upper side (step f), so that a very small aperture can be created. It is found that the protective Cr film formed at step d is not etched or damaged during the etching of the sample either in the TMAH or in the BHF. After forming the opaque layer, the actual size of the aperture is some-what reduced. With suitable metal coating, the effect of filling aperture is not found.

4.4.2   Glass process (see Figure 4.13)

For the capacitive-AFM/NSOM probe, the Si base is anodically bonded with the Pyrex glass. A narrow gap between the cantilever and the capacitive electrode is formed on the glass base. The Pyrex glass base of 2×2 cm^2 and 300 µm thick is used as a starting material (step a). Next a 100 nm thick Au/Cr film is sputtered on the glass surface, followed by the photolithog-raphy and Au/Cr etching (step b). Using this photoresist and Au/Cr pattern as a mask, the glass is etched in the BHF for about 1–3 µm to form a capacitive gap (step c). The Au/Cr mask is then removed and capacitive electrodes are formed on the glass base by Pt/Ti or ITO evaporation and followed by a lift-off process (step d). The thickness of the capacitive electrode is about 100 nm. Finally, the glass base is partly diced according to the shape of the probe before doing the anodic bonding process.

4.4.3   Bonding process (see also Figure 4.13)

The fabricated Si and glass parts are bonded together by anodic bonding. The Si wafer and glass base are aligned by an aligner and bonded at 300–400°C with a high voltage of 300–1000 V for 5–10 minutes. The glass base is not only used for forming the capacitive structure, but it also serves as a support base against the break of the probe during handling. In some cases, during anodic bonding, a strong electrostatic force between the Si cantilever and the capacitive electrode may cause cantilever bending. This problem can be avoided by utilizing a sacrificial layer, such as an array of polyimide columns in between as a protective layer. After bonding, this polyimide layer is removed in an oxygen plasma etching.

4.5   Fabrication results and discussion

With the described fabrication process, AFM/NSOM and capaci-tive–AFM/NSOM probes with apertures of diameter as small as 20 nm are successfully fabricated. Typical scanning electron microscopy (SEM) images of the fabricated SiO_2 tip with a focus on the apex of the Si cantilever are shown in Figure 4.14. It is observed that the shape of the SiO_2 tip is very well defined and its curvature angle at the apex area is very large (almost flat in

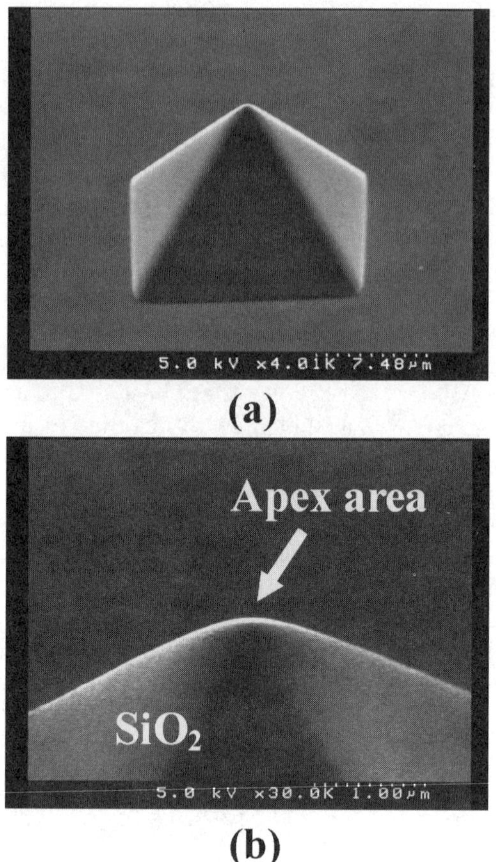

(a)

(b)

Figure 4.14 (a) SEM image of the fabricated SiO_2 tip on the Si cantilever before etching in the BHF (b) close-up view of the apex of the SiO_2 tip.

the area of about 0.5×0.5 μm^2 or nearly equal to the curvature of a sphere with about 1 μm in diameter). This is one of the essential features of the tip in terms of yielding a high optical throughput. Moreover, since the shape of the SiO_2 tip is determined by the initial shape of the pyramidal etch pit, the reproducibility of the tip's shape is very high compared to the optical fiber tip.

By etching the SiO_2 tip for the BHF for 6 min at 36°C, an approximately 500 nm aperture was successfully formed as shown in Figure 4.15a. After coating with an Al/Cr opaque film of 150 nm in thickness, the final NSOM apertured tip was formed as shown in Figure 4.15b. It is clearly observed that the aperture was successfully fabricated and located exactly at the apex of the SiO_2 tip without any alignment. Pinholes were not found in the side walls of the tip. The shape of the aperture is very well defined (mostly circular in shape). It is also an essential feature of the fabricated probe in terms of delivery of optical near-field light with high intensity. The size of the aperture at the apex of the SiO_2 tip is primarily determined by the

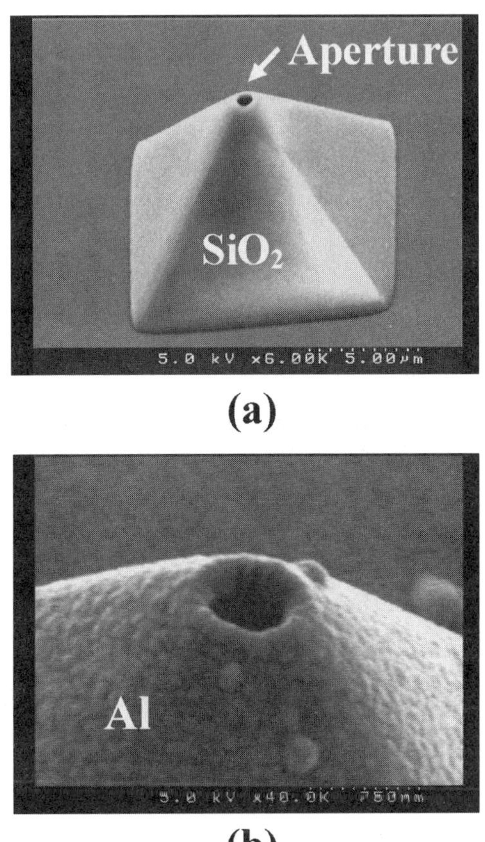

(a)

(b)

Figure 4.15 (a) SEM images of the SiO₂ tip with 500 nm aperture after etching in the BHF for 6 min at 36°C, (b) SEM image of the aperture after coating 150 nm Al/Cr opaque layer.

thickness of the SiO₂ and etching time in the BHF. Therefore, apertures with high uniformity in size were achieved. This is also an essential feature of the technique, not just for the NSOM probe, but also for application in near-field optical data storage where an aperture array is utilized.

By optimizing the etching time at a constant temperature, apertures with diameters as small as 20 nm or smaller (approximately 0.00001 of the diameter of a human hair) can be fabricated with the described technique. Experimental data shown in Figure 4.16 shows the dependence of the aperture size on the etching time in the BHF solution at 36°C. The deviation of aperture size with measuring 300 apertures on 2×2 cm² wafer is typically around 50 nm (± 25 nm). This deviation may due to an inhomogeneity of the SiO₂ grown at step c or the nonuniformity of the SiO₂ etching rate at step f. Moreover, the presence of the Cr protective layer may have some effect on the SiO₂ etching process. This deviation would be reduced by optimizing

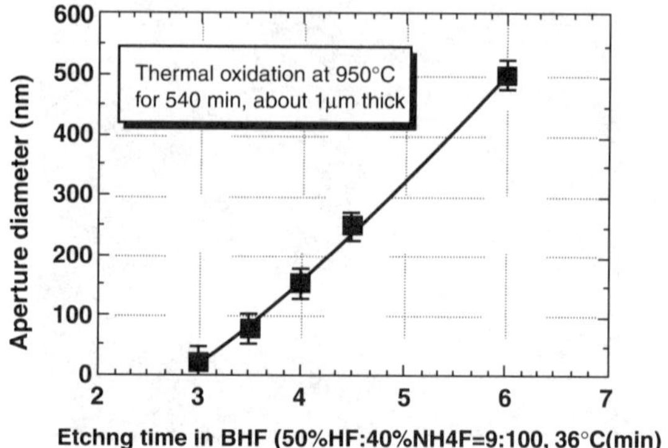

Figure 4.16 The dependence of the aperture size on the etching time in the BHF solution at 36°C.

the etching conditions. The described LOSE technique is advantageous in aperture fabrication for NSOM probe and has high reproducibility and high uniformity in mass production.

Since the key point of this technique is the etching of the SiO_2 in the BHF. The influence of the temperature and concentration of the etchant should be investigated. The dependence of the SiO_2 etching rate on the etching temperature in the BHF is shown in Figure 4.17. To precisely control the aperture size, the BHF-like etchant (e.g., $50\%HF:40\%NH_4F:H_2O = 9:100:X$) might be used to extend the SiO_2 etching time, which should be done at a constant, low temperature. The dependence of the SiO_2 etching rate on the concentration of H_2O in the BHF-like etchant is shown in Figure 4.18. Furthermore, during Si etching in the TMAH, the SiO_2 is also slightly etched. The Si and SiO_2 etching rate in the TMAH at 80°C were measured of 0.48 and 0.000151 μm/min, respectively,

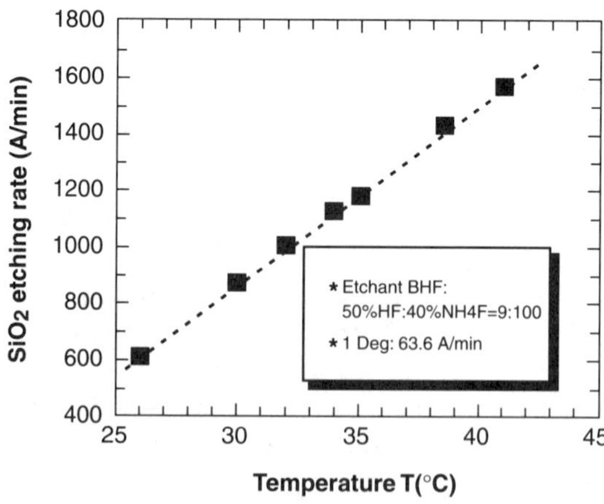

Figure 4.17 The dependence of the SiO_2 etching rate in the BHF on the etching temperature.

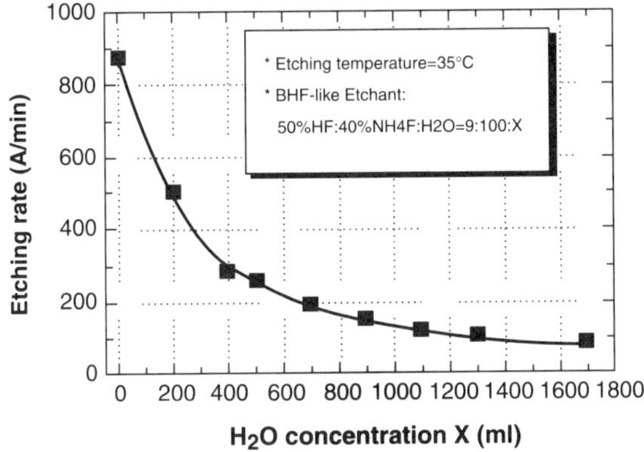

Figure 4.18 The dependence of the etching rate on the concentration of H_2O in the BHF-like etchant (50% HF:40%NH_4F:H_2O = 9:100:X).

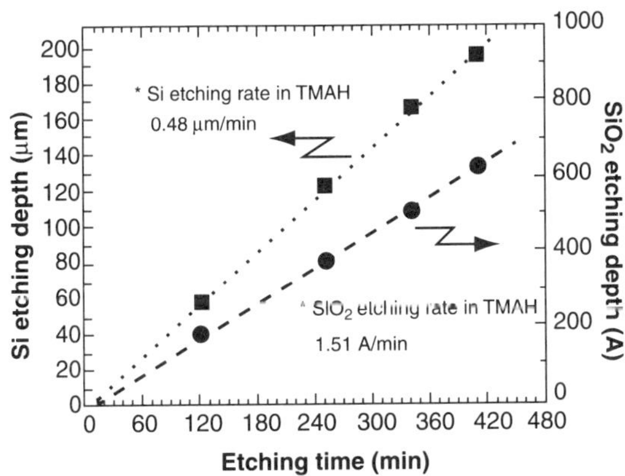

Figure 4.19 Si and SiO_2 etching rate in the TMAH etching at 80°C.

as shown in Figure 4.19. All of these parameters must be taken into account during the fabrication of the probe.

It should be noted that during formation of the opaque layer by evaporation or sputtering, even in case of an ultra-small aperture, the effect of filling of the aperture did not happen. It was experimentally observed that after coating with a thick opaque layer, the size of the metallic aperture is somewhat reduced and can be approximately expressed as $d = d_o - \alpha t$, where d is the diameter of the metallic aperture after metallization, d_o is the initial diameter of the aperture formed on the SiO_2 tip, and t is the thickness of the opaque layer, α is an experimental coefficient in between 0.5 and 0.7 (see Figure 4.20).

If the opening window for forming the etch pit in step b of the Si process (shown in Figure 4.12) is very square, four {111} Si surfaces will intersect at a point, therefore a single circular aperture can be created by etching as

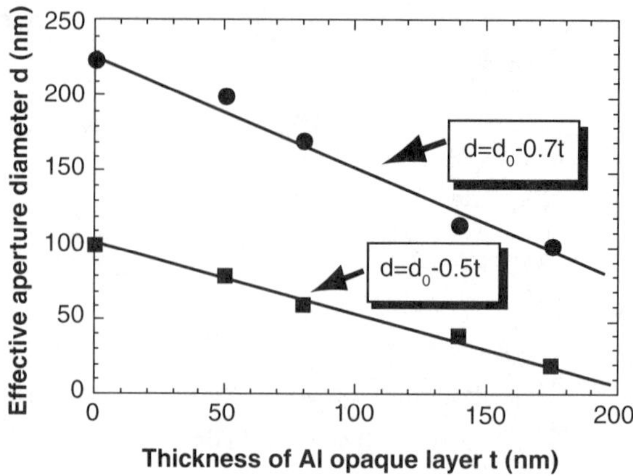

Figure 4.20 Aperture size as a function of thickness of the opaque layer (experimental data).

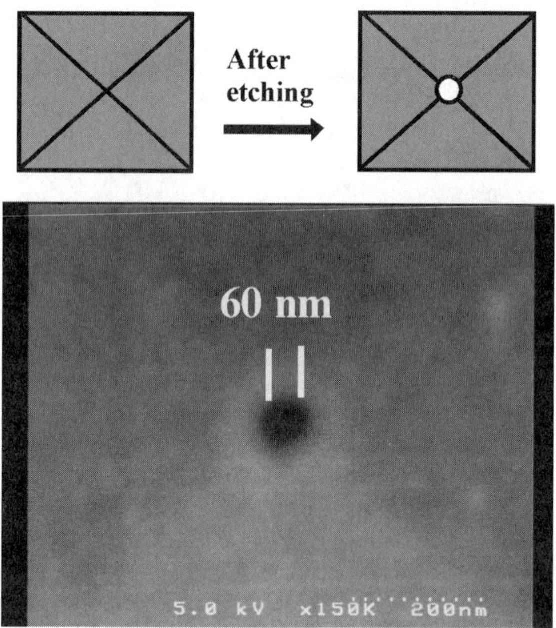

Figure 4.21 Forming of single aperture: SEM images of the fabricated 60 nm single aperture at the apex of the SiO$_2$ tip.

shown in Figure 4.21. If the opening window for forming the etch pit has a rectangular shape, a double aperture can be formed as shown in Figure 4.22.[32] Even two apertures on an SiO$_2$ channel can be fabricated as shown in Figure 4.23a. The distance between the centers of the two apertures is the same as the discrepancy in size of the opening window of the etch pit. Normally the

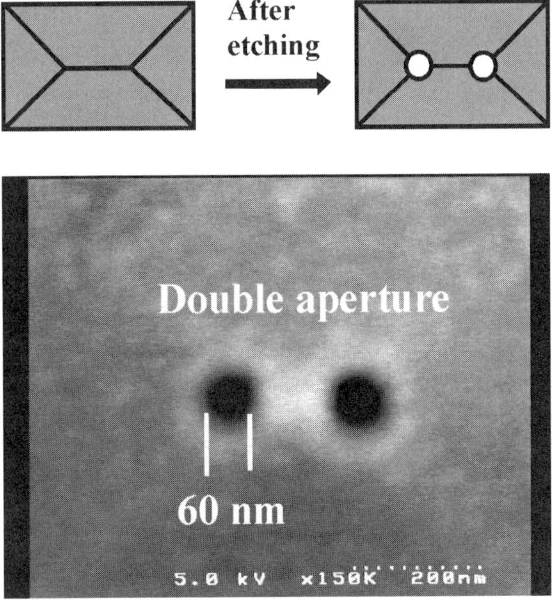

Figure 4.22 Forming of double aperture: SEM images of the fabricated 60 nm double aperture at the apex of the SiO$_2$ tip.

size of the two apertures in a double is identical. It is hoped that readers may find interesting applications with the double aperture. To fabricate only one small aperture, or double aperture with desired distance between the apertures, the patterning window for the etch pit should use high-resolution electron beam lithography.

With the described process, the aperture array at the apexes of the SiO$_2$ tip array, for instance, for recording application can also be fabricated as shown in Figure 4.23b.[33] Typical SEM images of the fabricated AFM/NSOM and capacitive–AFM/NSOM probes with apertures of 20–25 nm in diameter are shown in Figure 4.24, and Figure 4.25, respectively.

In summary, a well-defined aperture or aperture array with diameters as small as 20 nm (normally around 50–100 nm) were successfully fabricated in a simple batch process. The reproducibility both in size and shape of the tip and of the aperture is very high in comparison to the optical fiber based NSOM probes. The optical performance of the aperture tip and mechanical properties of the cantilever is much better than that of the optical fiber based NSOM probes. In the following chapters, optical performance and applications of the fabricated aperture and its array in near-field imaging and recording will be presented. Moreover, based on the fabrication process shown in Figure 4.12, several novel tips for locally enhancing the near-field intensity at the aperture and metallic contact in nanoscale for a thermal couple or heater can also be utilized. The process is also applicable in the fabrication of probes for microfield emitters, nano Auger.

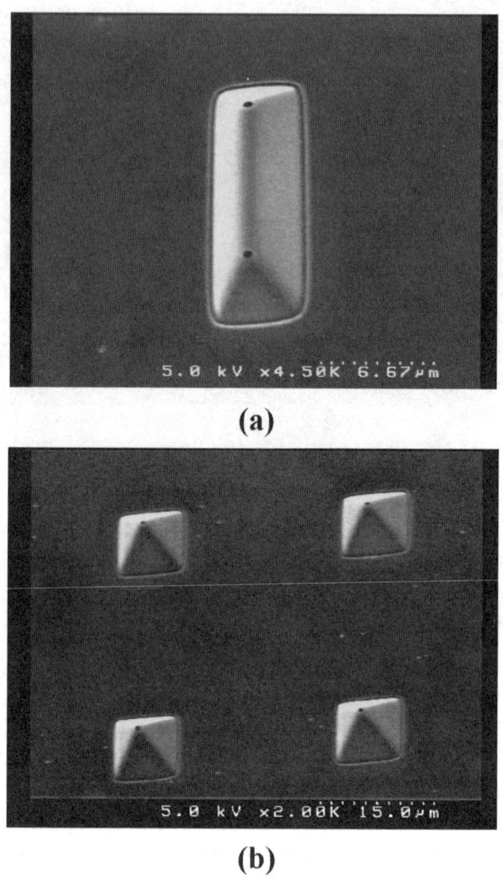

Figure 4.23 (a) Micro channel with two apertures at the apex of the SiO$_2$ tip; (b) SEM images of the fabricated aperture array at the apex of the SiO$_2$ tips.

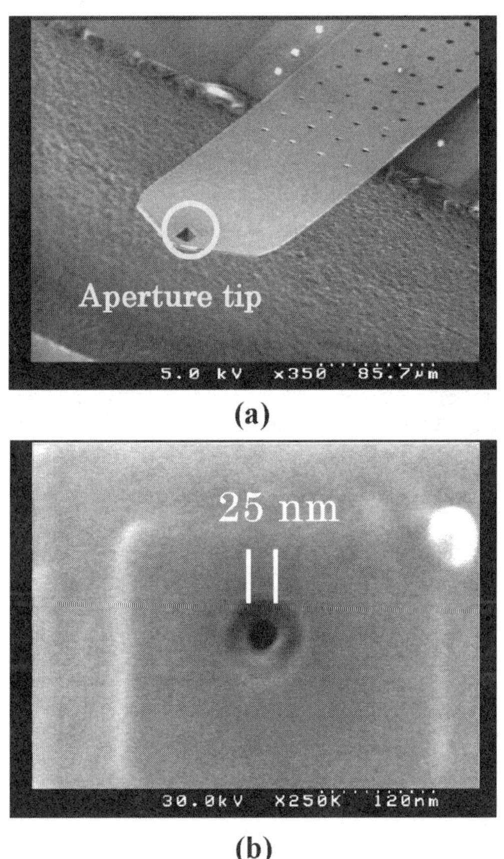

Figure 4.24 SEM images of the fabricated AFM/NSOM probe (a) and close-up view of 25 nm aperture at the apex of the tip (b).

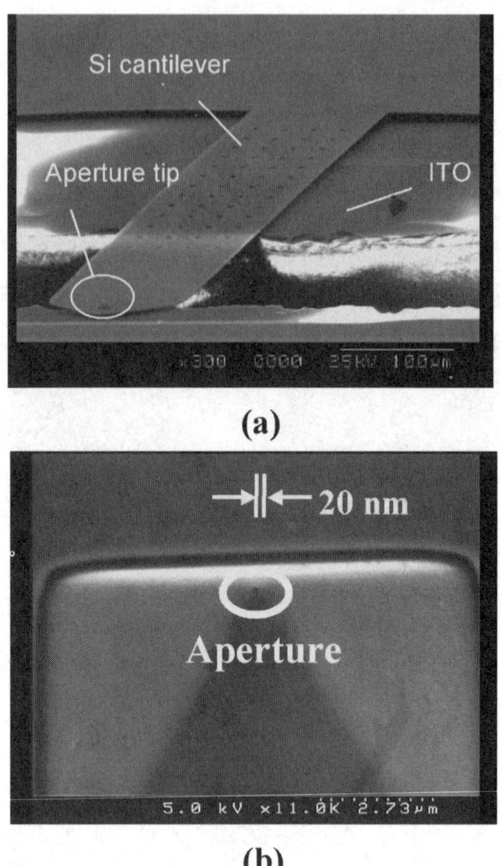

Figure 4.25 SEM images of the fabricated capacitive–AFM/NSOM probe (a) and close-up view of the tip with 20 nm aperture (b).

References

1. Moers, M.H.P. et al., Combined photon scanning tunneling microscope and atomic force microscope using silicon nitride probes, in *Near Field Optics*, Pohl, D.W. and Courjon, D., Eds., NATO ASI series, Kluwer, Dordrecht, 242, 79, 1993.
2. van Hulst, N.F. et al., Near-field optical microscope using a silicon-nitride probe, *Appl. Phys. Lett.*, 62, 461, 1993.
3. Ruiter, A.G.T. et al., Microfabrication of near-field optical probes, *J. Vac. Sci. Technol.*, B14, 597, 1996.
4. Oesterschulze, E., On the development and potential of cantilever-based probes for SNOM applications, *Opt. Mem. Neur. Net.*, 7, 251, 1998.
5. Mihalcea, C. et al., Multipurpose sensor tips for scanning near-field microscopy, *Appl. Phys. Lett.*, 68, 3531, 1996.

6. Werner, S. et al., Cantilever probes with aperture tips for polarization-sensitive scanning near-field optical microscopy, *Appl. Phys. A*, 66, S367, 1998.

7. Scholz, W. et al., in *Atomic Force Microscopy/Scanning Tunneling Microscopy 3*, Cohen, S.H. and Lightbody, M.L., Eds., Kluwer Academic/Plenum Publishers, New York, 1999, p. 75.

8. Prater, C.B. et al., Improved scanning ion-conductance microscope using microfabricated probes, *Rev. Sci., Instrum.*, 62, 2634, 1991.

9. Noell, W. et al., Micromachined aperture probe tip for multifunctional scanning probe microscopy, *Appl. Phys. Lett.*, 70, 1236, 1997.

10. Drew, D. et al., Nanostructured probes for scanning near-field optical microscopy, *Nanotechnology*, 10, 61, 1999.

11. Zhou, H. et al., Novel scanning near-field optical microscopy/atomic force microscope probes by combined micromachining and electron-beam nanolithography, *J. Vac. Sci. Technol.*, B17, 1954, 1999.

12. Akamine, S., Kuwano, H., and Yamada, H., Scanning optical microscope using an atomic force microscope cantilever with integrated photodiode, *Appl. Phys. Lett.*, 68, 579, 1996.

13. Minh, P. N., Ono, T., and Esashi, M., Fabrication of submicron aperture for near-field optical silicon probe, *National Conference on Physical Sensors*, Tokyo, 41, 1998.

14. Suzuki, Y. et al., Near-field aperture fabricated by solid-solid diffusion, *Appl. Phys. Lett.*, 77, 3710, 2000.

15. Schurmann, G. et al., Micromachined SPM probes with sub 100 nm features at tip apex, *Surf. Interface Anal.*, 27, 299, 1999.

16. Eckert, R. et al., Near-field fluorescent imaging with 32 nm resolution based on microfabricated cantilevered probes, *Appl. Phys. Lett.*, 77, 3695, 2000.

17. Mihalcea, C., Vollkopt, A., and Oesterschulze, E., Reproducible large area microfabrication of sub-100 nm apertures on hollow tips, *J. Elec. Soc.*, 147, 1970, 2000.

18. Albrecht, T.R. et al., Microfabrication of cantilever styli for the atomic force microscope, *J. Vac. Sci. Technol.*, A8, 3386, 1990.

19. Shiba, Y. et al., Capacitive AFM probe for high speed imaging, *Technical Digest of the 16th Sensor Symposium*, Tokyo, 269, 1998.

20. Akamine, S. et al., Microfabricated scanning tunneling microscope, *IEEE-Electron Device Lett.*, 10, 490, 1989.

21. Marcus, R.B. and Sheng, T.T., The oxidation of shaped silicon surfaces, *J. Elec. Soc.*, 129, 1278, 1982.

22. Hsueh C.H. and Evans, A.G., Oxidation induced stresses and some effects on the behavior of oxide films, *J. Appl. Phys.*, 54, 6672, 1983.

23. EerNisse, E.P., Viscous flow of thermal SiO_2, *Appl. Phys. Lett.*, 30, 290, 1977.

24. EerNisse, E.P., Stress in thermal SiO_2 during growth, *Appl. Phys. Lett.*, 35, 8, 1979.

25. Kao, D.B. et al., Two dimensional thermal oxidation of silicon-I: Experiments, *IEEE Trans. Elec. Dev.*, ED-34, 1008, 1987.

26. Kao, D.B. et al., Two dimensional thermal oxidation of silicon-II: Modelling stress effects in wet oxides, *IEEE Trans. Elec. Dev.*, ED-35, 25, 1988.

27. Marcus, R.B. et al., Formation of silicon tips with <1 nm radius, *Appl. Phys. Lett.*, 56, 236, 1990.

28. Albrecht, T.R. et al., Microfabrication of cantilever styli for the atomic force microscope, *J. Vac. Sci. Technol.*, A8, 3386, 1990.

29. Akamine, S. and Quate, C.F., Low temperature thermal oxidation sharpening of microcast tips, *J. Vac. Sci. Technol.*, B10, 2307, 1992.

30. Minh, P.N., Ono, T., and Esashi, M., Nonuniform silicon oxidation and application for the fabrication of aperture for near-field scanning optical microscopy, *Appl. Phys. Lett.*, 75, 4076, 1999.

31. Minh, P.N., Ono, T., and Esashi, M., Microfabrication of miniature aperture at the apex of SiO_2 tip on silicon cantilever for near-field scanning optical microscopy, *Sens. Actu.*, A80, 163, 2000.

32. Minh, P.N., Ono, T., and Esashi, M., High throughput aperture near-field scanning optical microscopy, *Rev. Scientif. Instr.*, 71, 3111, 2000.

33. Minh, P.N. et al., High throughput optical near-field aperture array for data storage, *Tech. Dig. of the 14th IEEE Int. Conf. on Mic. Elec. Mech. Sys.*, (*MEMS '01*), Interlaken, Switzerland, 309, 2001.

chapter five

Evaluation of microfabricated optical near-field probes

5.1 Optical throughput measurement

One of the challenges of the optical near-field technique, especially in the field of near-field data storage and spectroscopy, is how to increase the intensity of near-field light that is confined at the aperture or optical throughput. Optical throughput is defined as the ratio of the optical near-field power over the corresponding far-field power (or intensity). It is known that the optical throughput of an aperture NSOM probe is drastically decreased with decreasing the size of the aperture.

The optical throughput of the conventional optical fiber tip is quite low, e.g., 100 nm diameter aperture of optical fiber shows an approximately 10^{-5}–10^{-7} throughput.[1] In optical fiber based NSOM probes most of the light is lost at the tapered region due to the absorption by the metallic cladding and the cut-off effect that occurs at the fiber tip where its diameter is smaller than the wavelength of the incidence light.

To evaluate the optical throughput of the fabricated aperture NSOM probe, we proposed a simple measurement setup as shown in Figure 5.1. Detail of the measurement has been reported.[2] The fabricated probe is mounted on a stage of a conventional AFM system (Olympus-NV2000). A laser diode of 780 nm wavelength and 2mW power is focused on the tip area with an objective lens of 50x, NA = 0.55 (the laser diode was already mounted on the AFM system). A conventional photodiode with a sensitivity of 0.5 mA/mW was used as a detector to measure the near-field light passing through the aperture, or the far-field light without an aperture, and is placed on an XYZ piezo scanner of the AFM. The near-field or far-field light beam enters the photodiode and produces a photocurrent, which is proportional to the photon intensity. The photocurrent in the output of the photodiode is converted into a voltage signal and amplified by an electronic circuit. The

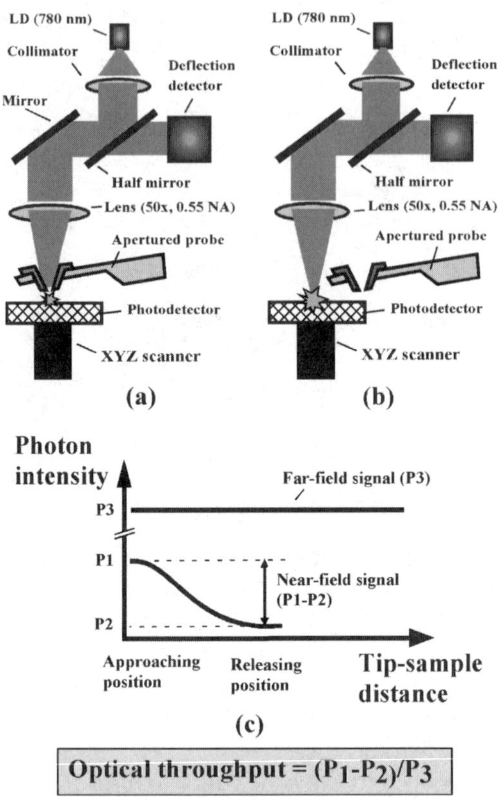

Figure 5.1 (a, b) Schematic diagram of the throughput measurement setup; (c) the near-field intensity is defined as the difference of the output signal at the approaching and releasing positions (P1–P2). The throughput is defined as (P1–P2)/P3.

near-field intensity is defined as the difference between the output signals at the position where the tip is almost in contact with the photodiode surface (the approaching position P1) and the releasing position (P2) as shown in Figures 5.1a and 5.1c. The distance between the approaching and releasing positions is several micrometers and is driven by the piezo scanner of the AFM. After evaluating the near-field intensity (P1–P2), the probe is taken away from the laser position without any change of the optical system, so that the far-field light beam directly illuminates on the photodiode (P3) as shown in Figure 5.1b. It was observed that the far-field signal P3 is nearly unchanged when adjusting the tip–sample distance between the approaching and releasing positions. The throughput of the corresponding aperture is defined as the ratio of (P1–P2)/P3. Using the above measurement setup and definitions, throughput of several apertures is evaluated and plotted as filled circles (see Figure 5.2). Note that the spot size of the laser beam is about 1.6 µm and the size of the tip is about 15×15 µm². There was no incident far-field light, therefore, outside the tip area that could enter the photodetector.

The optical transmission efficiency (or optical throughput) of a single hole with subwavelength diameter in an infinite metal film has been studied by Bethe using a scalar potential functional approach.[3] When the diameter of the hole (d) is smaller than the optical wavelength λ, the optical transmission scale as $(d/\lambda)^4$.[4] The optical throughput of a single hole with different sizes on the infinite metallic screen with different wavelengths, according to Bethe's model, was calculated and presented in Chapter 2 (Figure 2.9). For the case of an aperture on the SiO_2 tip, however, it is impossible to apply Bethe's model for evaluating the optical throughput because of the complexity of the geometry of the tip. We employed the finite difference time domain (FDTD) simulation for our aperture tip. Approximately 1–2% optical throughput for 100 nm diameter aperture was evaluated from the FDTD data (hollow square in Figure 5.2). Detail of the FDTD simulation will be addressed in Chapter 7. Furthermore, from the measurement of the spatial distribution of near-field light, an optical throughput of 2.4% for 300 nm aperture was estimated as indicated in Figure 5.2, hollow circle. This is also in agreement with the measurement. Detail of the spatial distribution measurement and optical throughput evaluation will be explained in the next section.

From the measurement result shown in Figure 5.2, it can be seen that the optical throughput of the fabricated probe is high. For example, for a 100 nm aperture, a throughput of 1% was achieved. This is much higher than that of the conventional optical fiber probe. The throughput of the probe is also higher than that of other types of micromachined tips.[5,6] The optical throughput shown in Figure 5.2 indicates that the micromachined probe has the potential to be used in many applications with localized near-field light.

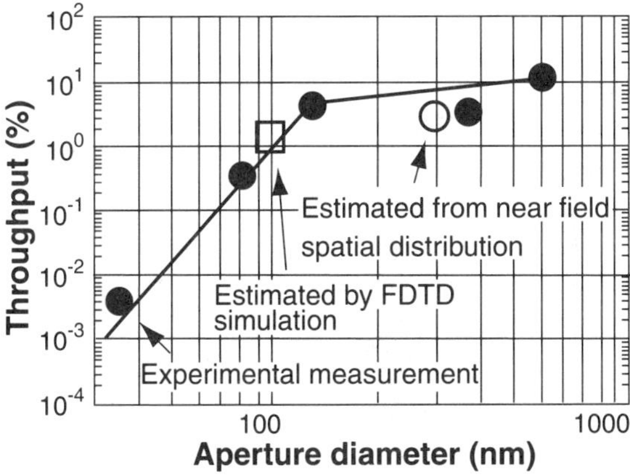

Figure 5.2 Throughput of the fabricated aperture NSOM probe as a function of the aperture diameter. Filled circle: experimental data; hollow square: estimated by FDTD simulation; hollow circle: estimated from the spatial distribution measurement.

The optical throughput of the NSOM probe is strongly related to the light propagation and absorption inside the tip and the cut-off effect. Systematic studies by Novotny et al.[7] and Hecht et al.[8] have proved that all the modes of light propagating in the optical fiber NSOM probe vanish when diameter of the tip become smaller than a critical value, i.e., a cut-off. At the cut-off the wave vector becomes imaginary and the mode changes from propagation to evanescent. A cross-section of the conventional optical fiber NSOM tip and optical mode analysis with incident light of 480 nm predicted by Hecht et al.[8] is presented in Figure 5.3a. It is shown that only the HE_{11} mode with lowest cut-off is able to go toward the aperture. Most of other modes were cut off and absorbed by the metallic cladding. That is the reason why optical throughput of the optical fiber based NSOM probe is very low.[2] A cross-section of the apertured SiO_2 tip is schematically shown in Figure 5.3b (which is based on the SEM image of the cross-section of the tip). Figure 5.3c shows cross-sections of the optical fiber and microfabricated tips having the same aperture size (100 nm) for comparison. It is clear that the taper angle of the micromachined tip is much larger than that of the optical fiber tip. For the microfabricated tip, the cut-off position is closer to the aperture resulting in higher optical throughput. It should be noted again that the shape of the fiber tip depends on the detail of the pulling or etching process and therefore is not very reproducible. The shape of the micromachined SiO_2 tip, however, is always reproducible.

5.2 *Measurement of spatial distribution of the near-field light*

The near-field light at the aperture does not propagate into space but locally situates at the proximity of the aperture. Measuring the spatial distribution of the near-field light at the aperture is a difficult task due to the weakness of the near-field signal and the difficulty in employing a nanoscale-sized detector close to the aperture. This difficulty was overcome by using the probe-to-probe measurement method developed by Ohtsu et al.[9,10] An experimental setup for measuring the spatial distribution of near-field light at the aperture is shown in Figure 5.4a and detail of the measurement was reported.[11-13] The microfabricated tip with approximately 300 nm aperture is located on a prism that can be moved by controlling an XYZ stage and a piezo-actuator. We scanned a nanosized light source emitted from an Au coated optical fiber probe with a 100 nm aperture and 10^{-5} optical throughput over the fabricated tip. Figure 5.4b shows the optical image of the head of the measurement setup. The near-field light emitted from the optical fiber tip is scattered by the micromachined aperture, transmitted through a prism, collected by a collective lens, and detected by a photomultiplier tube. The distance between the fiber tip and the fabricated aperture was maintained at 10 nm by a shear force feedback. The fiber was attached to a piezo-actuator and laterally vibrated at 300 kHz with an amplitude of 10 nm. Using the described experimental setup, the optical

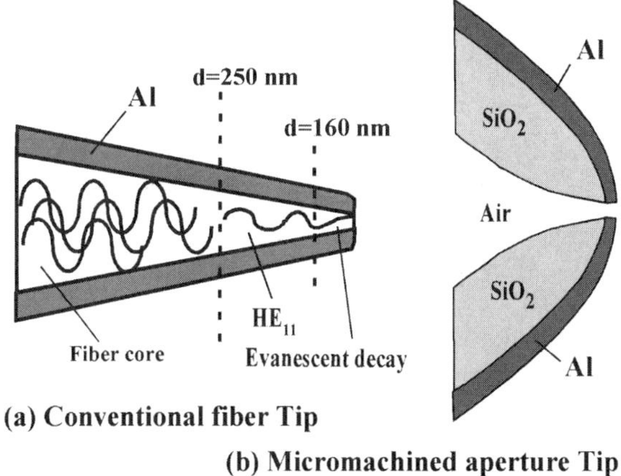

(a) Conventional fiber Tip

(b) Micromachined aperture Tip

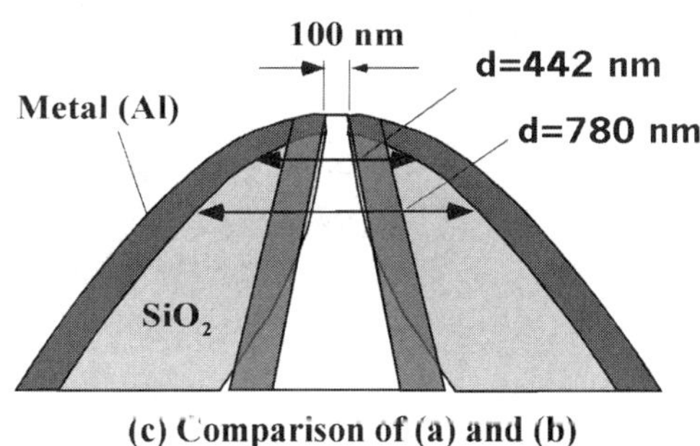

(c) Comparison of (a) and (b)

Figure 5.3 (a) Cross-section view and optical mode inside the conventional optical fiber NSOM tip (see Hecht 2000); (b) cross-section view of the microfabricated NSOM tip obtained from the FIB cutting; (c) comparison of (a) and (b) for 100 nm diameter tips.

near-field distribution was measured at a 300 nm aperture with an approximately 120 nm Al coating (Figure 5.5a). The measured near-field distribution is shown in Figure 5.5b. A cross-section of the near-field intensity at the center of the aperture (extracted from line AA in Figure 5.5b) is shown in Figure 5.6. The cross-section shows a full width at half maximum (FWHM) of 250 nm with a quite sharp peak of 60 nm wide. This means that a resolution of smaller than the size of the aperture can be achieved.

Based on the spatial distribution of the near-field distribution shown in Figure 5.6, we estimated an optical throughput of 2.4% for 300 nm aperture.[13] This value is close to the experimental result of the optical throughput measurement shown in Figure 5.2. We obtained an estimate of the optical throughput by comparing the height of the near-field intensity peak shown in Figure 5.6 with that of a 10% optical throughput near-field light emitted from an optical fiber probe with an approximately 1000 nm aperture. Strong confinement of near-field light at the micromachined tip and high optical throughput is likely related to the geometrical structure of the tip.[12]

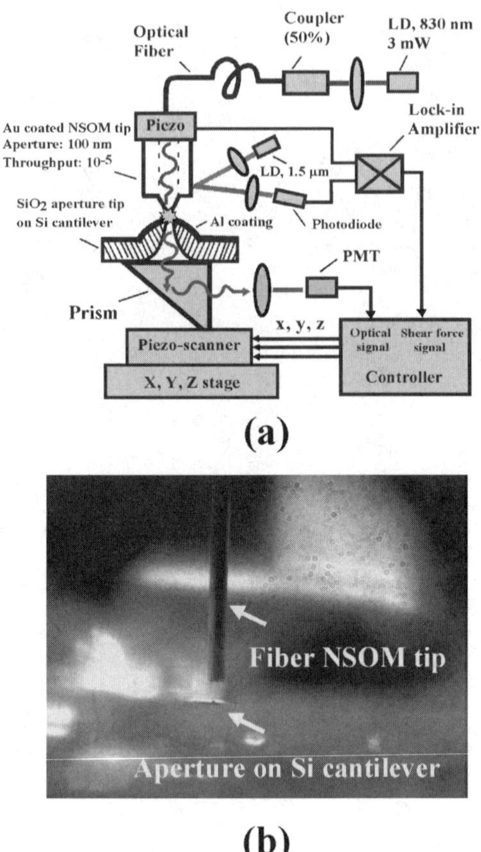

Figure 5.4 (a) Schematic diagram of the probe-to-probe method for measurement of the spatial distribution of near-field light at the aperture; (b) optical image of the measurement setup.

It should be mentioned that the diameter of the optical fiber used in this experiment was 100 nm. Therefore the near-field intensities of the micromachined tips with sub-100 nm aperture diameters were not determined. It was noted that the spatial distribution of near-field light at the aperture measured by the described setup (where the optical fiber is a light source) is identical for the case of apposite (where the optical fiber is a detector). This means that the spectrum shown in Figure 5.5b and Figure 5.6 can be reasonably considered as the spatial distribution of the fabricated aperture.

5.3   *Polarization behavior of the microfabricated aperture*

Polarization is a common property of electromagnetic waves. If the electric field vector of the light is at a random angle in all directions, the light is said to be unpolarized. Conversely, if the electric vector is forced along the same

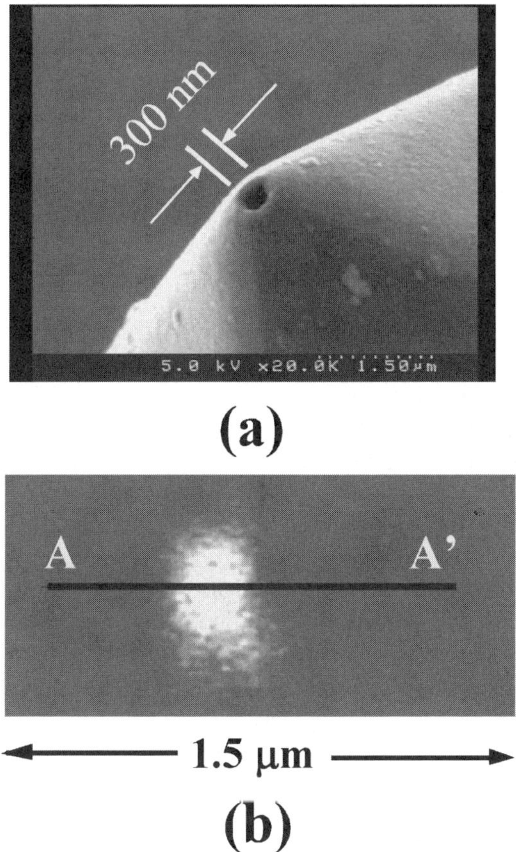

Figure 5.5 (a) SEM image of the 300 nm aperture used for measurement of optical near-field spatial distribution; (b) NSOM image showing the spatial distribution of near-field light at the aperture shown in (a).

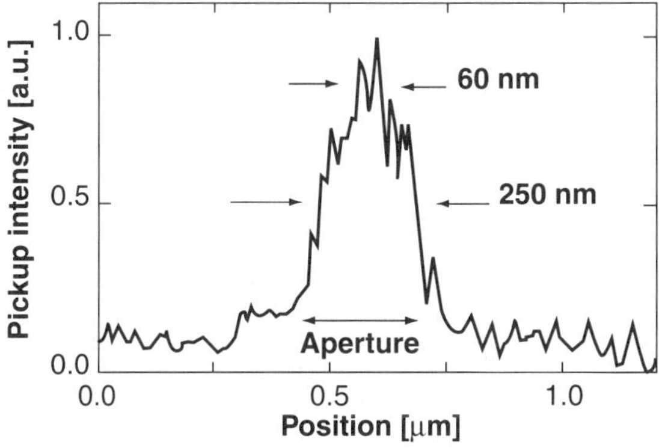

Figure 5.6 Cross-sectional view extracted from AA line in Figure 5.5b.

plane, the light is polarized. Polarization management is essential but diffi-cult in near-field optics. Normally for an optical fiber NSOM probe, it is difficult to control the polarization because the multireflection in the tapered region of the fiber and the difficulty in controlling the shape of the metallic coated aperture. Therefore polarization is normally arbitrary. To control the polarization a well reproducible shaped aperture formed by FIB cutting has been reported.[14]

The high reproducibility of the SiO_2 tip and aperture allows the polar-ization behavior of the microfabricated NSOM probes to be controllable. Moreover, the propagation length of the light inside the micromachined tip is very short compared to the optical fiber (see Section 5.1), the depolarizing effect due to multireflection is minimized. This section describes an experi-mental measurement of the polarization behavior of the fabricated apertured probe with different shapes.[11]

As described in Chapter 4, depending on the shape of the pyramidal etch pit formed on the Si wafer, one or two apertures can be fabricated at the apex of the SiO_2 tip. The distance between the two apertures is identical to the discrepancy in size of the opening window for forming the etch pit. To determine the dependence of polarization on the shape of the aperture, three types of apertures — rounded single aperture, connected double aper-ture and separated double aperture — have been fabricated as shown in Figure 5.7a, Figure 5.8a and Figure 5.9a, respectively.

The polarization behavior of photons emitted through these three kinds of apertures were analyzed in the far-field using the experimental setup shown in Figure 5.10. The microfabricated aperture is mounted on an XYZ stage of an optical transmission microscopy (Olympus-BX50). The white light from a halogen lamp is passed through an optical band path filter of 546 nm, a polarizer and a pinhole before focusing on the aperture. The far-field light passing through the aperture is collected by a lens (40x, 0.75 NA) and detected by a photon-counting camera (Hamamatsu-VIM). The photon intensity of the same area and same integrating time is counted for every polarizing angle. The polarization behavior of the fabricated apertures shown in Figure 5.7a, Figure 5.8a and Figure 5.9a are measured for every polarizing angle and plotted in Figure 5.7b, Figure 5.8b and Figure 5.9b, respectively. The circumpolar in the figure shows the direction of the electric field of the incident polarized light.

It is seen that for a rounded single aperture, due to the high symmetry of the tip and aperture, the detected photon intensity is nearly independent of the polarization angle (see Figure 5.7b). This is a useful property in the fabricated probe, not only for eliminating the artificial effect of the optical near-field imaging in a polarization–modulation NSOM that improves the delivery of the near-field light at the aperture, but also for improving the signal-to-noise ratio in, for instance, near-field data storage. The precision of Si micromachining technology allows the shape of the tip and aperture to be highly reproducible.

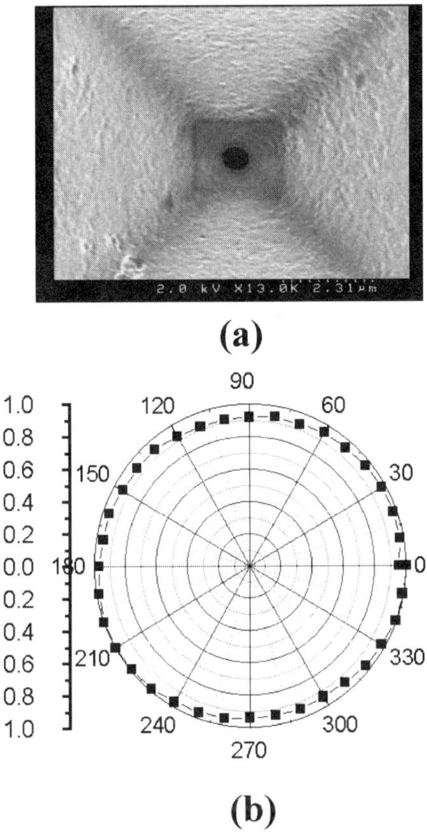

(a)

(b)

Figure 5.7 (a) SEM image of the rounded single aperture; (b) corresponding polarization behavior that was measured by the setup shown in Figure 5.10.

For the connected double aperture shown in Figure 5.8a, photon intensity is strongly dependent on polarization angle (see Figure 5.8b). The photon intensity is maximal at the position where the electric vector of the incident light is perpendicular to the long axis of the aperture. This type of aperture may be useful for near-field optical switching by changing the polarization angle of the incident far-field light. Applications for this type of aperture are under investigation.

For the separated double aperture in Figure 5.9a, photon intensity is also dependent on the polarization angle (see Figure 5.9b) even though the two separated apertures are very round. This means that the detected photon intensity may be affected by the coupling of the light between the two apertures and the interaction with the surface plasmon at the surface of the metallic bridge at the center of the tip. A simulation with the FDTD has shown that there is a strong enhancement of photon intensity at the center of the metal bridge. This could be developed as a structure for a high-resolution near-field tool even if the size of the aperture is large.

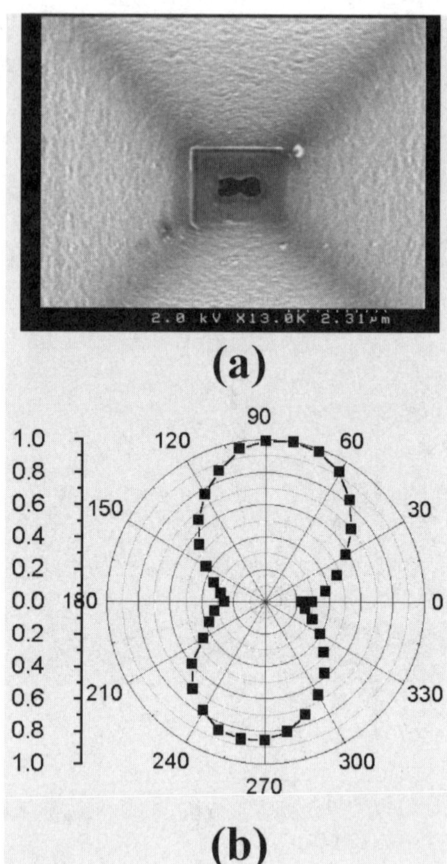

Figure 5.8 (a) SEM image of the connected double aperture; (b) corresponding polarization behavior that was measured by the setup shown in Figure 5.10.

Polarization behavior is still under investigation; however, these results suggest that the photon intensity with polarized light is strongly affected by the shape of the aperture. Since shape and size of the tip as well as the aperture are highly reproducible, the near-field polarization is controllable with microfabricated aperture NSOM probe.

5.4   Static and dynamic properties of fabricated cantilevers

In the capacitive–AFM/NSOM structure mentioned in Chapter 4, (Figure 4.7), when applying a DC voltage (V) between the cantilever and opposite electrode with a gap of d, an electrostatic force (F) is given by

$$F = 0.5\varepsilon_o S(V/d)^2 \qquad (5.1)$$

And the electrostatic force F_{end} applied at the end of the cantilever is given by

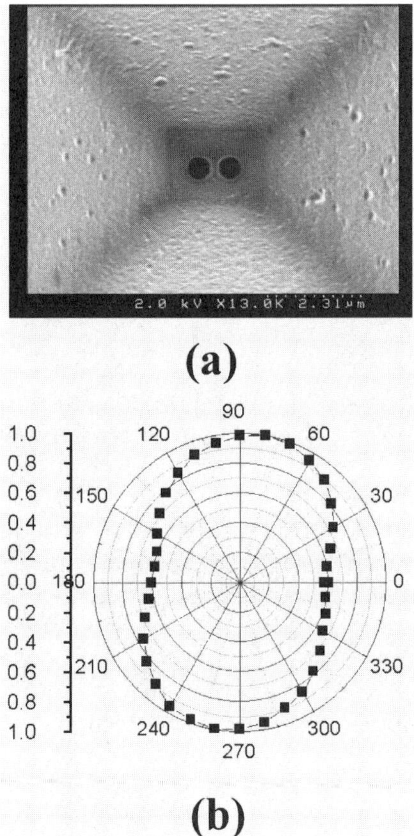

Figure 5.9 (a) SEM image of the separated double aperture; (b) corresponding polarization behavior that was measured by the setup shown in Figure 5.10.

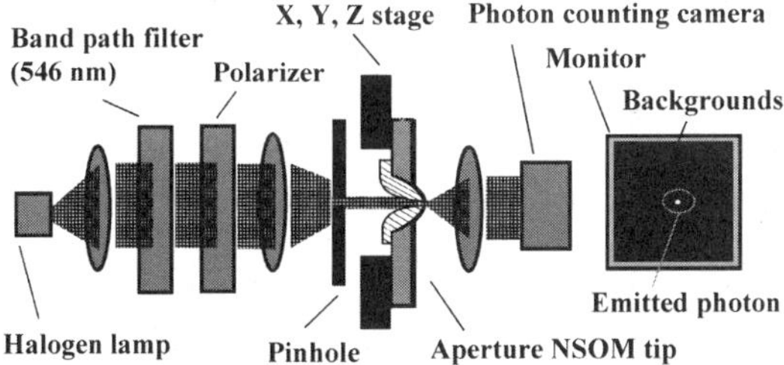

Figure 5.10 Experimental setup for measurement of the polarization behavior of the fabricated aperture using the photon counting camera (Hamamatsu-VIM).

$$F_{end} = 0.5\varepsilon_o S_{eff}(V/d)^2 \tag{5.2}$$

where S is the area of the capacitor (S = L.w); S_{eff} is effective area (S_{eff} = 3wL/8); ε_o is the electrical constant of air. This electrostatic force causes cantilever bending with a static deflection z and is given by

$$z = (1/6)\varepsilon_o(S_{eff}/d^2)\ (L^3/EI)V^2 \tag{5.3}$$

where I is the geometrical moment of inertia; E is the Young module of silicon. The capacitance between the cantilever and the opposite electrode is given as follows:

$$C = \varepsilon_o\ S_{eff}/d \tag{5.4}$$

Using the capacitive–AFM/NSOM probe with 1 μm capacitive gap, 0.5 μm thick, 80 μm wide and 100 μm long the mechanical properties (both static and dynamic modes) of the probe were investigated. The deflection of the cantilever when applying voltage is measured with an optical position sensor. The measured and calculated results of the dependence of the cantilever deflection and the applied voltage are shown in Figure 5.11. The dependence of the measured capacitance on the applied voltage is shown in Figure 5.12. It is seen that the cantilever can be deflected about 0.9 μm by applying a DC voltage of 5V.

The dynamic characteristic of the probe was measured in a vacuum of 4×10^{-2} Pa using a Doppler laser vibration detection system (NEOARK MLD-102-1Z Nihon Kagaku Eng.). A simplified illustration of the measurement

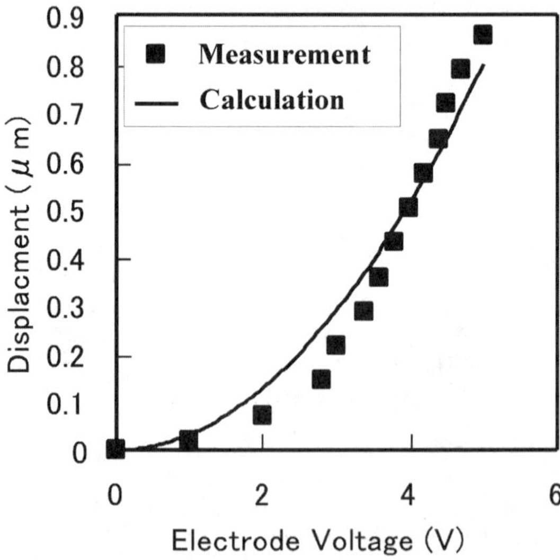

Figure 5.11 Relationship between the deflection of the cantilever and the applied DC voltage. The deflection was measured with an optical position sensor when a DC voltage was applied.

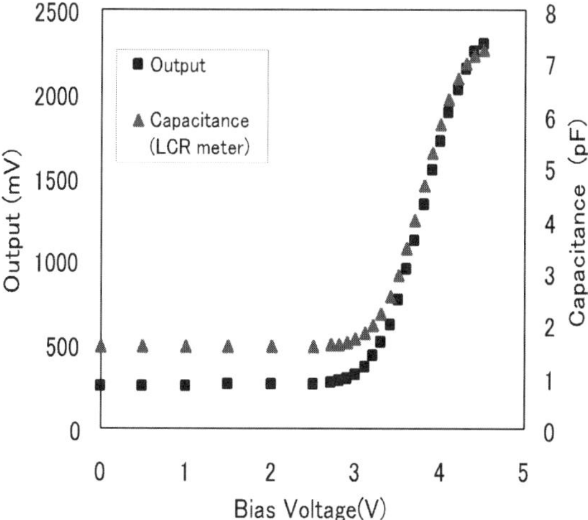

Figure 5.12 Relationship between the measured capacitance of the probe and the applied DC voltage.

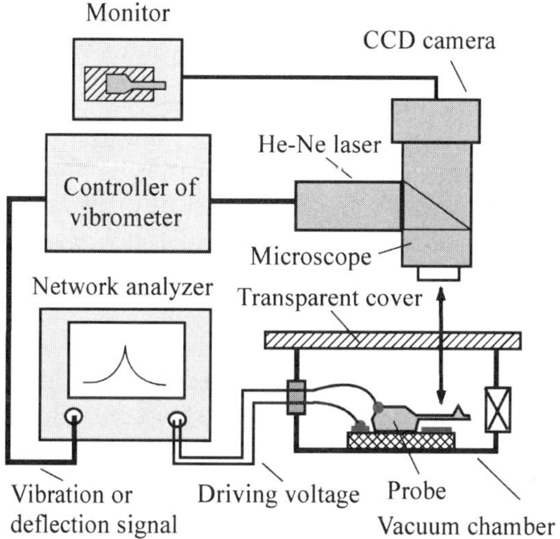

Figure 5.13 Schematic diagram of the Doppler laser vibration detection measurement.

setup is shown in Figure 5.13. The probe is located in the vacuum chamber. A driving voltage is applied to the probe from the output of a network analyzer (Advantest: R3751B). The vibration of the cantilever is monitored by the Doppler laser system. The amplitude and phase of the vibration signal of the cantilever are recorded with the network analyzer. Figure 5.14 shows the measured amplitude and the phase response with a resonant frequency

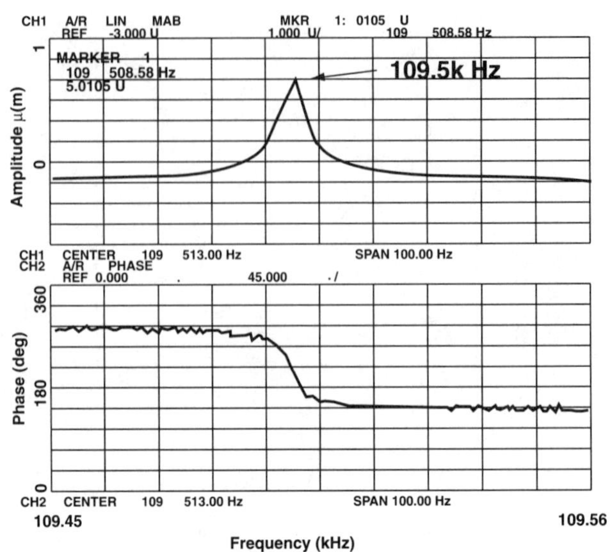

Figure 5.14 Resonance characteristic of the probe measured in 4×10^{-2} Pa.

of approximately 109 kHz, and Q factor is 2×10^4. It is shown that even if the driven signal was small (3 mV), a large vibration amplitude (1 µm) of the probe was obtained because of the high Q factor in vacuum.[15] Nevertheless, due to air damping and mechanical energy loss, the cantilever hardly vibrated in air. The Q factor of the cantilever measured very low in air.

In summary, the capacitive–AFM/NSOM probe has functions for both sensing and actuation because the deflection of the microcantilever can be measured from the capacitance and it can be electrostatically actuated.

5.5 Discussion

Several essential optical and mechanical properties of the microfabricated probes were investigated. The most valuable features of the fabricated probe with SiO_2 tip are the high optical throughput, well-confined near-field light, controllable polarization and good mechanical properties. It was shown by several studies that the tip geometry is highly important because it governs two crucial aspects of near-field application — lateral resolution and optical throughput.[1,16-19] The important role of the tip geometry can be understood with, for example, the theoretical work done by Novotny and Pohl.[18] They showed that the optical throughput of a conical, metal-coated aperture probe can be improved by as much as 10,000 times with increasing the haft-cone angle from 15° to 30°, without having to increase the spot size.

It is impressive that the geometry of the fabricated tip is reproducible. The metal screen was formed on the outer side of the SiO_2 tip with a large opening angle (nearly equal to the curvature of a sphere with about 1 µm diameter). The laser light beam comes to the aperture without any strong losses inside the tip. It is reasonable to assume that increasing the opening angle will increase transmission because the cut-off region will be shortened. The gradual change in curvature angle of the inner and outer curvatures, as well as the metal screen on the outer side of the tip, may cause an additional effect similar to a convergent lens. Furthermore the physical aperture is empty and is applicable not only to photons but for other particles, such as electrons or ions as well.

Since the near-field image is strongly dependent on polarization, a controllable polarization light source is essential in near-field optics, which makes these microfabricated NSOM tips advantageous.

Nevertheless, for an aperture smaller than 100 nm, the optical throughput is still drastically decreased as shown in Figure 5.2. New types of apertures for improving optical throughput, such as coaxial or surface plasmon at the metallic particle located at the center of the aperture, are proposed and described in the next chapter.

References

1. Valascovic, G.A., Holton, M., and Morrison, G.H., Parameter control, characterization, and optimization of optical fiber near-field probes, *Appl. Opt.*, 34, 1215, 1995.
2. Minh, P.N., Ono, T., and Esashi, M., High throughput aperture near-field scanning optical microscopy, *Rev. Scientif. Instr.*, 71, 3111, 2000.
3. Bethe, H.A., Theory of diffraction by small holes, *Phys. Rev.*, 66, 163, 1944.
4. Ebbesen, T.W. et al., Extraordinary optical transmission through subwavelength hole arrays, *Nature*, 391, 667, 1998.
5. Zhou, H. et al., Novel scanning near-field optical microscopy/atomic force microscope probes by combined micromaching and electron-beam nanolithography, *J. Vac. Sci. Technol.*, B17, 1954, 1999.
6. Lee, M.B. et al., Nanometric aperture arrays fabricated by wet and dry etching of silicon for near-field optical storage application, *J. Vac. Sci. Technol.*, B17, 2462, 1999.
7. Novotny, L. and Hafner, C., Light propagation in a cylindrical waveguide with a complex, metallic, dielectric function, *Phys. Rev. E*, 50, 4094, 1994.
8. Hecht, B. et al., Scanning near-field optical microscopy with aperture probes: fundamentals and applications, *J. Chem. Phys.*, 112, 7761, 2000.
9. Kourogi, M. et al., Subwavelength sized phase change recording with a high throughput fiber probe, in *Near-field Optics: Physics, Devices and Information Processing*, Jutamulia, S., Ohtsu, M., Eds., *Proc. SPIE*, 3791, 68, 1999.
10. Yatsui, T., Kourogi, M., and Ohtsu, M., Highly efficient excitation of optical near-field on an apertured fiber probe with an asymmetric structure, *Appl. Phys. Lett.*, 71, 1756, 1997.
11. Minh, P.N. et al., Spatial distribution and polarization dependence of the optical near-field in a silicon microfabricated probe, *J. Micros.*, 202, 28, 2001.
12. Minh, P.N. et al., Near-field optical apertured tip and modified structures for local field enhancement, *Appl. Opt.*, 40, 2479, 2001.

13. Minh, P.N. et al., Near-field distribution at the aperture of the microma-chined NSOM probe, *61st Autumn Meeting of the Jap. Soc. Appl. Phys.*, Sapporo, Japan, 896, 2000.

14. Veerman, J.A. et al., High definition aperture probes for near-field optical microscopy fabricated by focused ion beam milling, *Appl. Phys. Lett.*, 72, 3115, 1998.

15. Shiba, Y. et al., Capacitive AFM probe for high speed imaging, *Tech. Dig. of the 16th Sensors Symposium*, 269, 1998.

16. Betzig, E. et al., Breaking the diffraction barrier: optical microscopy on a nanometric scale, *Science*, 251, 1468, 1991.

17. Ohtsu, M., *Near-field Nano/Atom Optics and Technology*, Springer-Verlag, Tokyo, 1998, p.45.

18. Novotny, L. and Pohl, D.W., Scanning near-field optical probe with ultra small spot size, *Opt. Lett.*, 20, 970, 1995.

19. Wiesendanger, R. and Guntherodt, H.J., Eds., *Scanning Tunneling Microscopy II*, Springer-Verlag, Heidelberg, 28, 323, 1995.

chapter six

Novel probes for locally enhancing near-field light and other applications

One of the most important features in near-field optics is the total amount of near-field light that can be coupled through an extremely small aperture. As mentioned previously, optical throughput of these microfabricated aperture probes is very high due to the geometric structure of the SiO_2 tip. However, optical throughput drastically decreases with decreasing aperture size, especially for apertures smaller than 100 nm (see Figure 5.2). To extend this application into other areas optical throughput needs to be improved.

The optical throughput of other types of NSOM probes, such as a coaxial NSOM with the aperture having a carbon nanotube at the center or an aperture with an embedded Ag particle, were examined. Even though a final result for such novel tips has not yet been achieved, a technological solution seems possible. A novel structure of a hybrid optical fiber-apertured cantilever is also presented here. An extension of the fabrication technique for a metallic contact in nanosize for a thermal profiling or thermal recording probe as well as for electron field emission devices has also been developed and presented in this chapter.

6.1 Fabrication of the coaxial apertured probes

Coaxial structure is well known in electronics as a non-cutoff waveguide. Using coaxial structure for high throughput NSOM has been proposed by Keilmann[1] and Fee et al.[2] to take the non-cutoff property of the structure. However, fabrication of such structures is still a challenge. To enhance the optical throughput of optical fiber based NSOM probes, several attempts have been made, such as using a coaxial structure with a thin metal wire at the center of the aperture,[3] or an aperture with a metallized protrusion at the center of the glass core.[4] A quite effective method has been developed where a double or triple tapered steps fiber probes or unsymmetrical shaped fiber tips were utilized to efficiently excite and couple the HE_{11} mode into the aperture area.[5,6] Other types of near-field probes using a metal microslit were proposed by Bae et al.,[7] where the slit had a width much smaller than that of the wavelength and a length of larger than $\lambda/2$. This was formed on the core of the optical fiber. Since the tapered slit serves as a simple parallel plate transmission line, cut-off effect does not occur. Oesterschulze et al. developed a coaxial aperture probe where a thin tungsten wire was grown at the center of the aperture by focused ion beam deposition for application in infrared spectroscopy.[8] These methods still suffer from the difficulty of the fabrication in a batch process. Another approach to fabricating a coaxial aperture NSOM probe in a batch process developed from the LOSE fabrication technique is proposed here.

When the protective Cr film on the upper side of the tip is thick enough (typically thicker than 150 nm), the etching rate of the SiO_2 in the apex area becomes faster than in other parts of the tip (see Figure 6.1a). The mechanism of this effect is unclear. It is possible that when the protective Cr protrusion starts to appear in the BHF etchant, the SiO_2 around the Cr tip is etched at a higher speed; this leads to the structure having a Cr protrusion at the center as shown in Figure 6.1a. By completely etching the SiO_2 tip, the shape of the protective Cr film can be observed. Figure 6.1b clearly shows the modification of the interior contour of the etch pit with low temperature oxidation. The SiO_2 etching rate is probably highest at the interface of the Cr/SiO_2. It was observed that the protective Cr layer on the upper side of the tip (formed by sputtering at 300 W, 25 mTorr of Ar at room temperature) causes tensile stress. It is presumed that this tensile stress is strongest at the apex area due to the geometrical structure of the tip. Since the binding between the Cr and the SiO_2 film is quite strong, the SiO_2 tip at the apex area develops some tensile stress thereby enhancing the SiO_2 etching rate in the BHF solution. Investigating the effect of enhancing the SiO_2 etching rate at the apex area of the tip would be an interesting matter. However, this study focuses on what kinds of applications could use this effect.

The results shown in Figure 6.1 suggest a novel structure of coaxial aperture for high throughput NSOM probes. Principles of the coaxial NSOM are sketched in Figure 6.2. The coaxial probe consists of a small aperture and a metallic nanowire at the center of the aperture. The fabrication process is shown schematically in Figure 6.3. First, the Si cantilever with an integrated SiO_2 tip that has a Cr protective layer is fabricated (step a) using the Si process described in Chapter 4 (Figure 4.12). Next, the aperture is formed at the SiO_2 tip by etching the SiO_2 tip in a buffered-HF solution (step b). As mentioned previously, tensile stress may be due to SiO_2 in the apex area being etched at a higher rate, (see Figure 6.1a). Next, a 100 nm thick opaque layer is formed by oblique evaporation of Al or Cr onto the lower side of the SiO_2 tip (step c). Finally, the Cr film on the upper side of the SiO_2 tip is obliquely etched (step d) with fast atom beam (FAB) etching. A wire-like Cr structure at the center of the tip is formed to serve as the metallic core of the coaxial tip. The metallic wire and the opaque layer are isolated by the

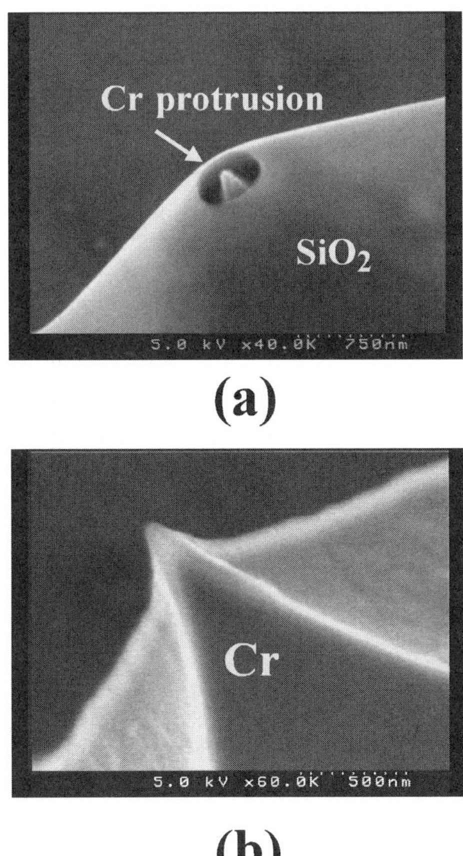

(a)

(b)

Figure 6.1 (a) SEM image of an SiO$_2$ tip with 150-nm thick Cr protective layer, after etching in BHF for 5 min. The aperture is created with a Cr protrusion at the center of the aperture; (b) SEM image of the Cr protrusion after completely removing the SiO$_2$.

SiO$_2$ tip and air at the aperture. SEM images of the proposed coaxial NSOM structure and its cross-sectional view before FAB etching is shown in Figure 6.4. The cross-section (Figure 6.4b) was formed by focused ion beam (FIB) milling. Note that to protect the SiO$_2$ tip and eliminate the charging effect during milling, a thick ITO was deposited as indicated in Figure 6.4b. This cross-sectional image was also used for building a model for simulation that will be explained in the next chapter.

Using the process shown in Figure 6.3, the coaxial NSOM tip was fabricated. Figures 6.5a and 6.5b, respectively, show front and back views of the tip taken with SEM of the initial fabrication of coaxial NSOM probe. With a suitably oblique angle during FAB etching, a Cr tip, which is suspended by four Cr wires can be formed at the center of the aperture as shown in Figure 6.5c (the image is taken from the upper side of the tip).

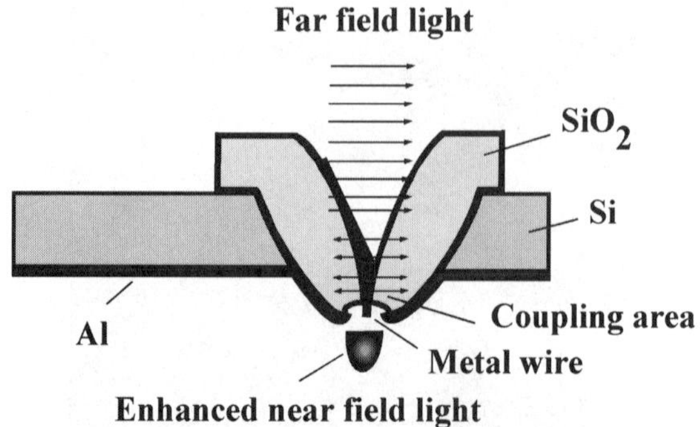

Figure 6.2 Principle of a coaxial aperture NSOM tip. The tip consists of a metallic wire-like metal at the center of the aperture of the SiO$_2$ tip.

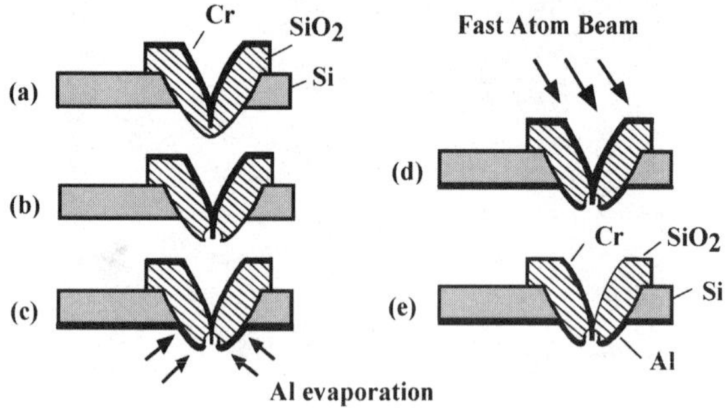

Figure 6.3 Fabrication process of a coaxial aperture NSOM tip.

The optical enhancement of this structure has not yet characterized, however, this structure seems to be a promising candidate for high through-put NSOM probes for several reasons. First, the coaxial structure could possibly be used to excite the TEM$_{00}$ mode because it has the unique property of focusing electromagnetic energy of the incident light.[1-3] Second, incident light could excite and be coupled with surface plasmon at the tip of the metallic nanowire, so that a strong enhancement of the near-field intensity at the aperture is expected.[9] Third, the asymmetric structure of the probe is also expected to excite the other light mode.[6] The enhancement of the near-field light of the coaxial structure compared to the non-coaxial structure was confirmed by a finite difference time domain (FDTD) simulation.

6.2   *Fabrication of the apertured probe with a single carbon nanotube*

A carbon nanotube (CNT) is a novel nanostructure consisting of one or several graphen sheets rolled into a tube.[10] Recently CNT has become a very

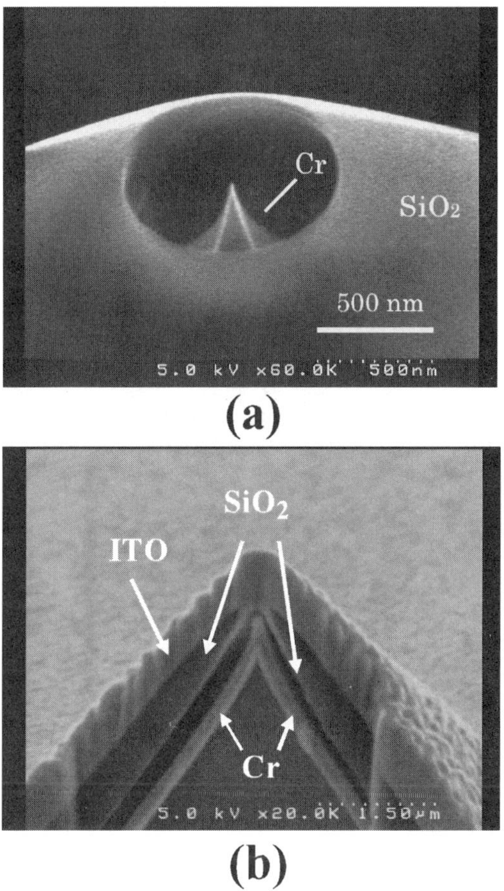

Figure 6.4 (a) SEM image of a tip before FAB etching process (b) cross section of the coaxial structure formed by FIB milling.

interesting development due to the mechanical, electrical and optical properties associated with a nanotube. CNTs normally have a high aspect ratio. Many applications with CNT have already been proposed and demonstrated.[11] The CNT as a super tip in the SPM probe has been successfully used as ultra-high-resolution atomic[12,13] and magnetic[14] force imaging. The CNT-based SPM probes have become an essential technique, not only for surface characterization, but also for the manipulation of small objects and probe-based high-density data storage.[15] Moreover, high-aspect ratios of the CNT together with its high chemical stability are advantageous for application to field emitters. These useful features enable the CNT based field-emission display.[16]

The CNT has been produced in outflow of carbon arc,[17] in laser vaporization of a graphite rod in an oven,[18] and in thermal decomposition of hydrocarbon using chemical vapor deposition.[19] It is required for many

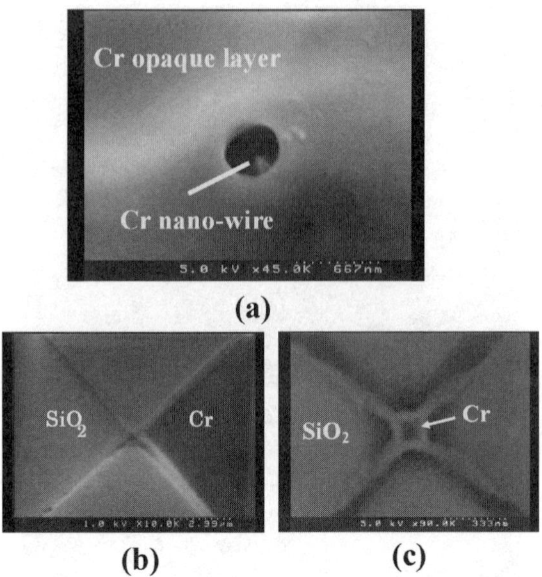

Figure 6.5 (a) SEM images of the first fabricated coaxial-NSOM with a wire-like Cr at the center of the aperture taken from the lower side; (b) wire-like Cr at the center of the aperture; (c) Cr tip at the center of the aperture, which is suspended by four Cr wires.

applications that desired CNTs are selectively grown on the desired position. The selective growth is ensured by the chemical vapor deposition method,[20] where CNT is grown on a catalyst metal such as Fe, Ni or Co.

  A simple apparatus for growing CNT has been developed using a hot-filament chemical vapor deposition (CVD). The hot-filament CVD apparatus was originally built for CNT growth as shown in Figure 6.6. The chamber

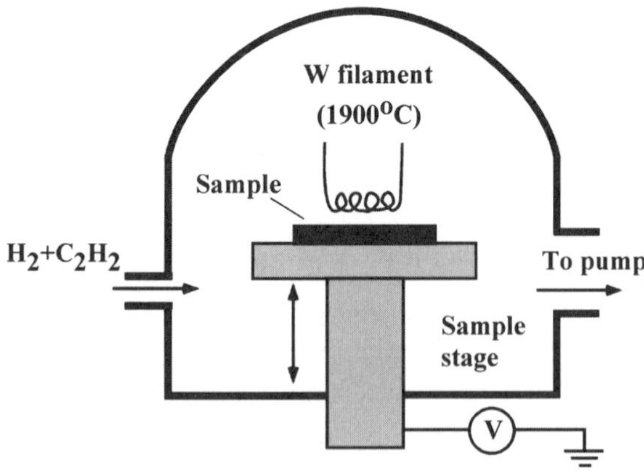

Figure 6.6 Schematic diagram of thermal chemical vapor deposition for growing carbon nanotube.

releases exhaust by a turbomolecular pump at 200 liters/sec. Tungsten filament with a diameter of 0.6 mm was placed above the sample stage. Axetylene (C_2H_2) gas diluted with hydrogen (H_2) gas at 27 Pa pressure was introduced into the growth chamber, which was thermally dissociated and activated on the heated filament at 1900°C. An optional voltage can be applied to the sample stage. In all experiments, CNT growth was performed for 15 min and under an enhanced electric field of 300 V. Details of the CNT growing technique are reported in reference 21.

Using the described apparatus, carbon nanotubes have been successfully grown selectively. For example, Figures 6.7a and 6.7b show the SEM images of the grown CNT in the quartz glass substrate and patterned Ni dots, respectively. Even a single CNT has been successfully grown at the apex of a commercial Si tip (AFM tip) using the electric field enhancement effect (see Figure 6.8a). Using the CNT grown AFM tip, atomic force image of SiC dot patterns has been successfully observed as shown in Figure 6.8b.[21] As a comparison, an AFM image of the same sample recorded with a commercial super-sharp Si tip is shown in Figure 6.8c. Clear differences in the morphology in Figures 6.8b and 6.8c can be observed. The parallelogram shape of the dots in Figure 6.8c shows tip artifacts. The image recorded with grown CNT tip is close to the real sample.

The observed effect shown in Figure 6.1 and the development of the CNT growing technique have suggested other technological approaches for fabrication of an aperture with a single CNT for NSOM and other applications. The fabrication process as shown in Figure 6.9 was developed for making aperture probes with a single CNT at the center of the aperture. As usual, the Si cantilever with the thermal SiO_2 tip was formed (step a). Next, a thin Ni or Fe film of 5–10 nm is sputtered on the upper side of the tip (step b). A transparent Si_3N_4 film about 500 nm thick is then deposited on the tip by sputtering or CVD to form a supporting layer (step c). After etching the SiO_2 tip in the BHF, the aperture with an Si_3N_4-supported Ni or Fe protrusion is formed at the center of the aperture (step d). Next, an Al film for opaque layer is deposited by oblique evaporation (step e). Finally, the wafer is transferred into the CNT growth chamber to grow a single CNT at the center of the tip (step f). A typical SEM image of the growing CNT at the center of the aperture is shown in Figure 6.10b. A transmission electron microscopy (TEM) study of the grown CNT proved the existence of the Ni particle at the top of the tube as shown in Figure 6.10c.

This structure is expected to serve as a high throughput NSOM by utilizing the surface plasmon enhancement of the Ni particle. Furthermore, CNT can be doped to form a metallic-like tube. In that case, the structure can be considered as a lightning rod and an ideal coaxial structure to suppress the loss of light propagation into the aperture. The CNT is also utilized to increase the lateral resolution in topographic imaging. With the fabrication process in Figure 6.9, it is feasible to fabricate an array of aperture tips having CNT for electron field emission applications as well.

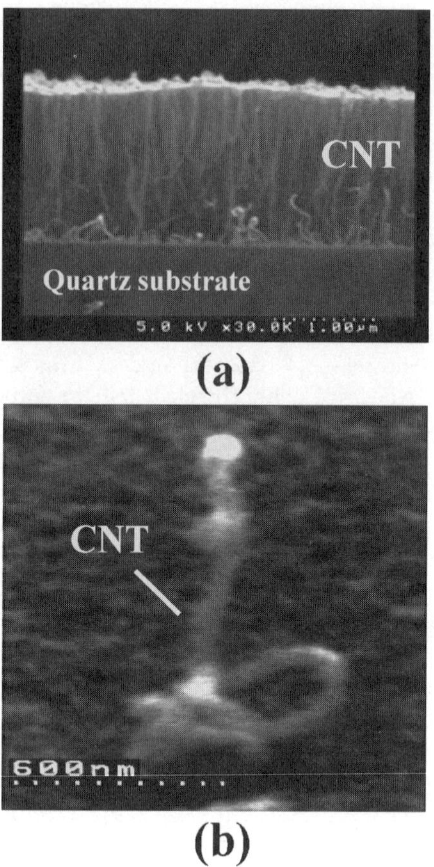

Figure 6.7 SEM images of the grown CNT in the quartz glass substrate (a) and patterned Ni dots (b) using the thermal CVD equipment.

6.3 *Fabrication of the apertured probe with an embedded Ag particle*

One of the effective ways to enhance the optical throughput in near-field optics is to use an Ag or Au particle as a probe. Such small Ag particles exhibit a localized plasmon resonance invisible, which greatly enhances the optical scattering cross-section. Surface plasmon is an electromagnetic wave with the collective oscillations of electrons that occur on the surface of a thin metallic film. To utilize the surface plasmon effect in the aperture NSOM probe, isolation of metal particles and the opaque metallic film is needed. Several attempts have been made to fabricate such a structure. For instance, Spalli et al. developed a technique to attach a small gold particle at the center of the aperture of the optical fiber tip. An enhancement both of the far-field and near-field light was observed.[22] Okamoto et al.[23] proposed a technique to attach a single Au particle at the apex of an SiN tip using an epoxy resin.

Figure 6.8 (a) SEM images of the grown CNT at the apex of a commercial Si tip using the electric-field enhancement; (b) AFM image of the SiC patterns recorded with the CNT grown AFM tip; AFM image of the SiC patterns recorded with the super-sharp commercial AFM tip.

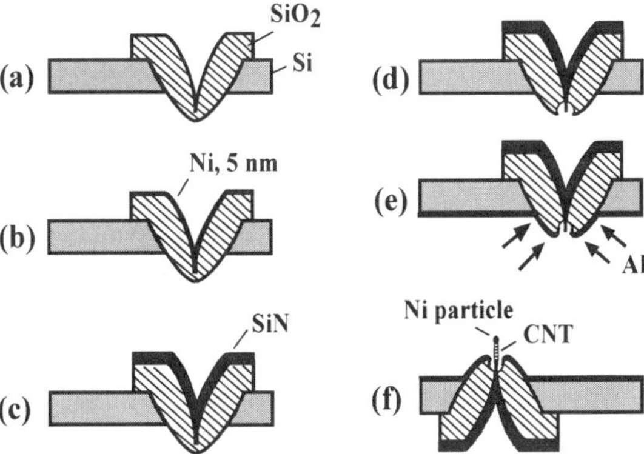

Figure 6.9 Fabrication process of a novel NSOM probe with a small aperture and a single carbon nanotube at the center of the aperture for enhancing the optical throughput and lateral resolution.

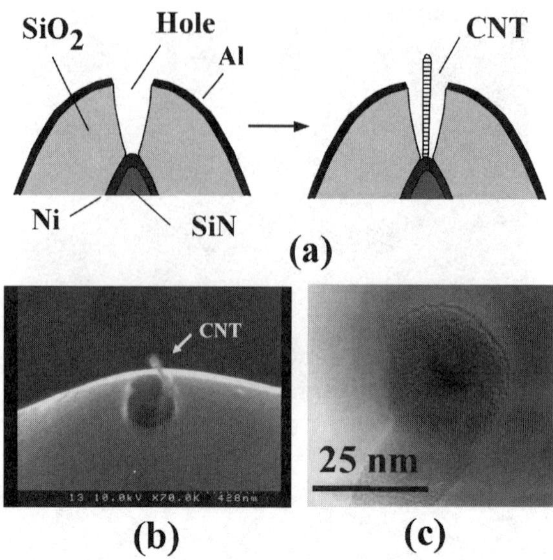

Figure 6.10 (a) Schematic principle of the growing of CNT at the center of the SiO$_2$ aperture; (b) SEM image of the apertured tip having a CNT at the center of the aperture; (c) transmission electron microscopy (TEM) image of the grown CNT with an Ni particle at its end.

The Au nanoparticle acts as a scattering center to effectively convert the evanescent light on the sample into the propagation light in the PSTM configuration. These techniques, however, suffer from difficulties in the fabrication process and are time-consuming. In this section another technique for fabrication of the aperture probe with Ag particle for high throughput NSOM applications in a batch process as presented.

According to the pioneering works done by Tominaga et al.,[24] small Ag particles can be formed on Ag$_2$O thin film by heating the Ag$_2$O film at a temperature up to 160–200°C according to the following relation:

$$Ag_2O - (T > 160°C) - \rightarrow Ag + 1/2O_2 \qquad (6.1)$$

The above reaction can be reversible in a closed system where oxygen exists. In an open system the reaction is usually irreversible. The effect was effectively applied for a super-resolution near-field system (Super-RENS) near-field recording where the enhanced near-field light by the Ag particle is utilized for writing or reading bits.

This result and the effect shown in Figure 6.1 was applied in the fabrication of a novel high throughput NSOM probe. The fabrication process is almost identical to the process of apertured CNT and shown in Figure 6.11. First, the Si cantilever with an SiO$_2$ tip is fabricated (step a). Next, a transparent 10–20 nm thick Ag$_2$O is formed on the upper side of the SiO$_2$ tip by sputtering with an Ag target in 0.5 sccm (cubic centimeter per minute at

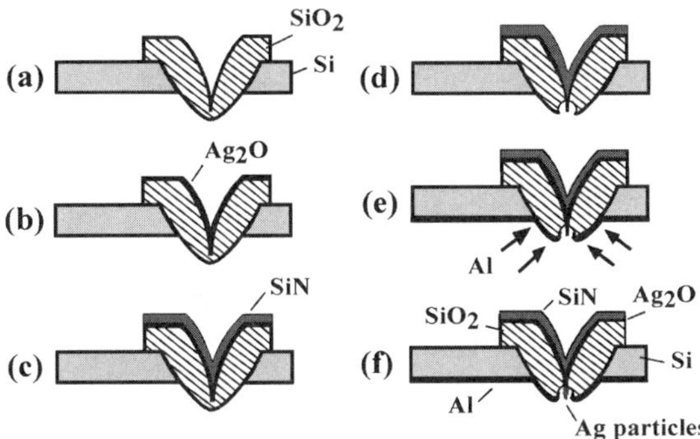

Figure 6.11 Fabrication process of a coaxial aperture NSOM probe with a small aperture and Ag particles at the center of the aperture for high throughput NSOM probe.

standard temperature and pressure) O_2 and 1.5 sccm Ar at a pressure around 1 Pa (step b). Under this sputtering condition, the Ag_2O phase is dominant.[24] Consequently, an approximately 500 nm thick Si_3N_4 film was formed on the Ag_2O film by sputtering (step c). The wafer is next dipped into the BHF for forming an aperture at the apex of the tip and the Si_3N_4-supported Ag_2O protrusion at the center of the aperture (step d). Next, an approximately 65 nm thick Al film is deposited on the outer side of the SiO_2 tip by oblique evaporation (step e). Finally, the sample is heated for 2 min, at approximately 200°C. After heating, Ag particles were formed at the apex of the Si_3N_4 tip because Ag_2O decomposed into Ag and O_2 according to the Equation (6.1). At the apex of the tip, oxygen can escape through the aperture leaving only Ag particles, whereas in the sidewall of the tip, Ag particles and oxygen can react to form the Ag_2O because of the protection by the Si_3N_4 and SiO_2 layers. The formation of Ag particles with 50–100 nm diameters have been observed on the Ag_2O/SiO_2 substrate by heating and the mechanism was confirmed by several other studies.[24-26] If this structure is realized, in illumination mode, the near-field light is expected to excite and couple with surface plasmon on the surface of the Ag particles so that a significant enhancement of the near-field intensity could be expected. In the collection mode, Ag particles can serve as a scattering center with high cross-section and the aperture suppresses the background far-field light. An SEM image of the initial fabrication of a surface plasmon NSOM tip with the described technique is shown in Figure 6.12. A close-up view of the aperture area of the surface plasmon NSOM tip in Figure 6.12b shows two or three Ag particles with diameters between approximately 50 and 60 nm at the Si_3N_4 tip.

It is very difficult to confirm by measurement that these are Ag particles because the area is quite small. However, the formation of Ag particles on other planar surfaces of SiO_2 under the same experimental conditions proved the formation of Ag particles. An array of aperture tips having Ag particles can be fabricated in a batch process for application in, for example, optical

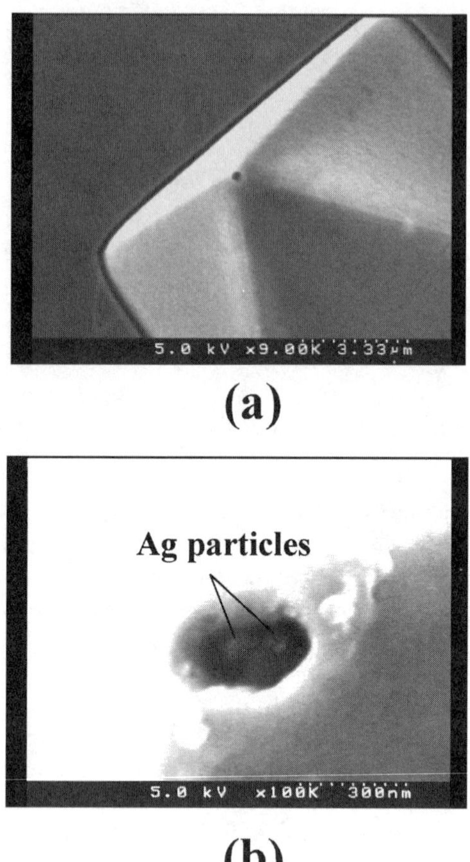

Figure 6.12 (a) SEM image of a fabricated aperture probe with Ag particles at the center of an aperture formed by the process shown in Figure 6.11; (b) close-up view of an aperture having two or three Ag particles.

near-field recording. Optical near-field distribution and application of the fabricated surface plasmon NSOM probes in near-field recording are under investigation.

6.4 *Fabrication and characterization of a hybrid structure of an optical fiber and cantilever*

Near-field optics still relies heavily on optical fiber based probes because there are no others readily available. The optical fiber based probes have an advantage in coupling and guiding the light from the far-field light source into the aperture or from the aperture to the detector. If the optical fiber is combined with the micromachined aperture, a better probe will be produced by using the advantage and eliminating the disadvantages of both the optical fiber and the micromachined aperture.[27,28] In this section, a new type of probe is proposed, a hybrid optical-fiber or fiber-bundle and micromachined apertures for parallel near-field imaging, processing, photolithography or data storage.

The schematic structure of the hybrid optical fiber-apertured cantilever is shown in Figure 6.13. Two types of probes were proposed, hybrid fiber-apertured SiO_2 cantilever and hybrid fiber-apertured SiO_2 diaphragm. This configuration makes the optical setup easy because light can be introduced from the other end of the optical fiber through the fiber itself. The flexible cantilever structure can be deflected in the normal direction to the cantilever plane, which enables the optical image to be in contact with a sample without the destruction of the probe. As mentioned in Chapter 5, because the opening angle of the SiO_2 tip is very large, the cut-off effect is minimized and the optical throughput of the hybrid probe is greatly improved. Moreover, the far-field light that is reflected back by the cantilever's tip can be utilized to monitor the deflection or vibration of the cantilever, i.e., to monitor the tip–sample distance by the probe itself.

The fabrication process of the hybrid optical fiber-cantilever NSOM probe is shown in Figure 6.14 (dimensions are not to scale) and has been described in detail in references 29-31. First, a 200 μm thick, (100) oriented Si wafer is oxidized, and the silicon dioxide is etched in the BHF using a resist mask formed by photolithography. Etch pits are formed by anisotropic etching of silicon in the TMAH (step 1). After removing the oxide, silicon oxide is thermally grown by wet oxidation at 950°C to the thickness of about 1 μm.

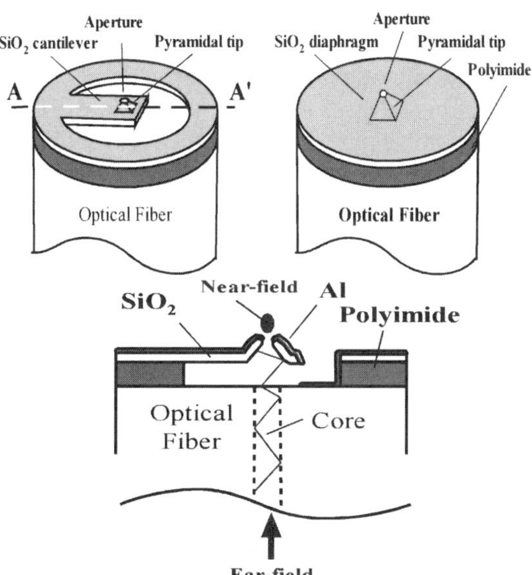

Figure 6.13 Schematic structures of a hybrid optical fiber-apertured cantilever and apertured diaphragm and its cross section.

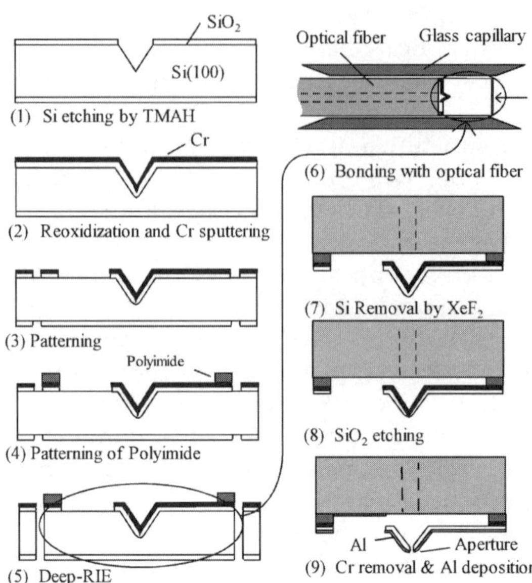

Figure 6.14 Fabrication process of a hybrid optical fiber-aperture cantilever or diaphragm.

Subsequently, a Cr film approximately 100 nm thick was sputtered (step 2) and patterned together with underlying silicon dioxide. The silicon dioxide on the front side of the wafer is patterned for the cantilever structure (step 3). An adhesive spacer with a thickness of about 5 μm was formed with a photosensitive polyimide (step 4). As many as 3000 silicon columns of 125 μm diameter having the cantilever patterns are formed on a 2×2 cm² Si wafer by deep reactive ion etching (deep-RIE) of silicon from the back side using the silicon dioxide as a mask (step 5). The Si column and the optical fiber are then inserted into a glass capillary with a diameter of 127 μm, pressed from both sides and bonded together with the adhesive polyimide spacer by heating at 360°C (step 6). Details of the assembling and bonding process have been described in references 32-33. After bonding, the combination optical fiber-Si column is taken away and the silicon column is etched out by isotropic etching with an XeF₂ gas (step 7). XeF₂ was chosen because of its outstanding selectivity in etching of the Si and the SiO₂. To fabricate a minute aperture, the fiber end is dipped into the BHF solution for etching the SiO₂ tip (step 8). Finally, the protective Cr layer on the upper side of the tip is etched out and an approximately 100 nm thick Al layer is deposited at the end of the fiber by evaporation to form an opaque layer (step 9).

Optical images of the fabricated Si column and the combination of optical fiber and Si column after bonding in the glass capillary are shown in Figure 6.15. SEM images of the fabricated probe with apertured cantilever and diaphragm and a magnified view of an aperture tip are shown in Figure 6.16. The diameter of the cladding and the core of the single-mode optical

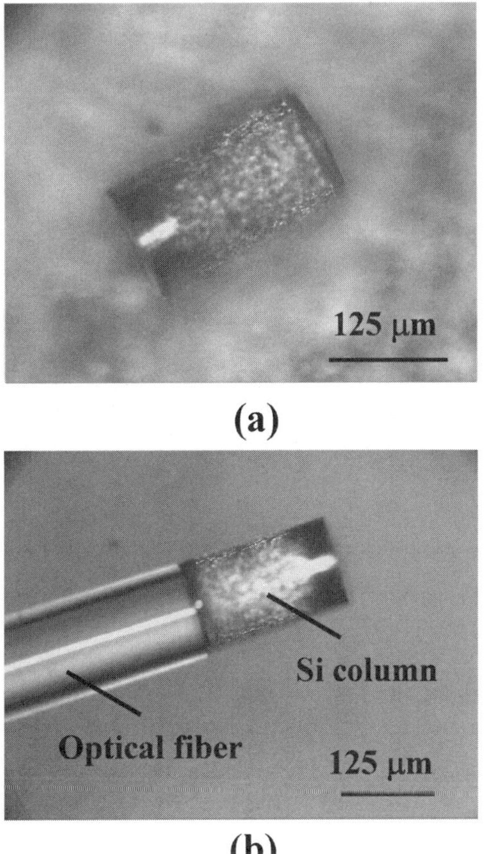

Figure 6.15 Optical images of a fabricated Si column (a) and optical fiber and Si column after bonding in the glass capillary (b).

fiber are 125 and 5 µm, respectively. The width and length of the cantilever beam are 35 and 60 µm, respectively. The size of the tip is 20×20 µm^2.

It should be noted that the alignment of the probe tip and the fiber core during the bonding process is essential because the light-coupling efficiency from the fiber to the aperture is determined by the accuracy of the alignment. To reduce misalignment during bonding, the diameter of the glass capillary should be close to the diameter of the optical fiber and Si column.

To improve the light coupling efficiency, a microlens should be placed in between the aperture and the fiber core. For the hybrid single fiber-micromachined aperture, a single glass ball lens can be put on the upper side of the etch pit before assembling with the optical fiber. A schematic illustration in Figure 6.17 shows the process of putting a glass ball lens on the upper side of the etch pit. A sharp metal tip is used to attach a glass ball lens on the surface of an Si substrate by the electrostatic force under a microscope, and put it into the pyramidal tip before bonding with the fiber

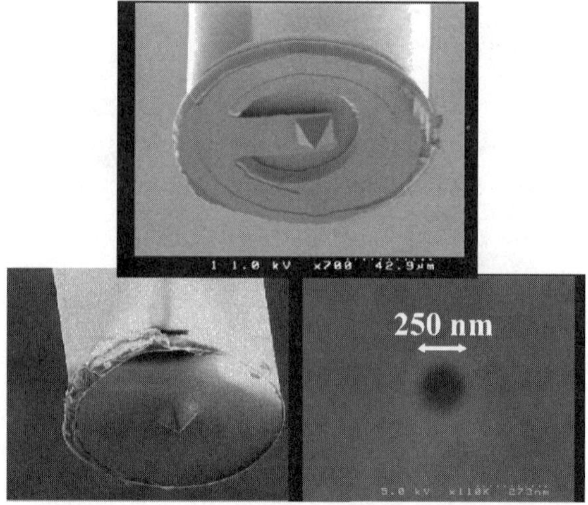

Figure 6.16 Typical SEM images of a fabricated hybrid optical fiber-apertured cantilever, apertured diaphragm and a close-up view of the aperture at the apex of the tip.

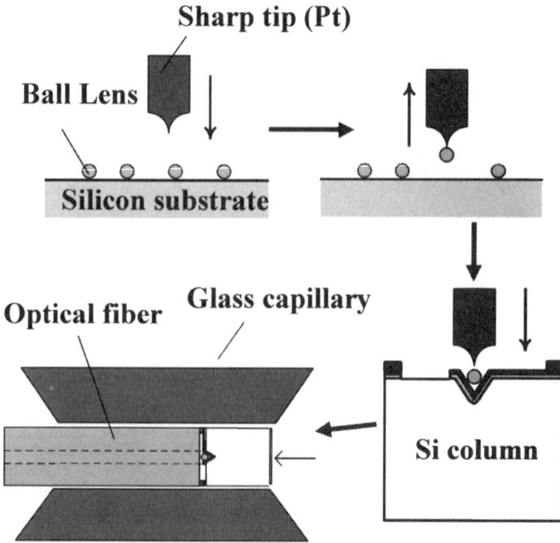

Figure 6.17 Schematic illustration of the manipulation method using a sharp metal tip for putting a glass ball lens on the upper side of the tip.

as shown in the figure. Figure 6.18a shows the optical image of the glass ball lens attached on the Pt tip according to the electrostatic force. Figure 6.18b shows the cantilever probe on the silicon column, where a glass ball lens with a diameter of 14 μm was inserted into the pyramidal tip before the assembling process. An output light from the end of the fiber can be collected by the ball lens and focused on the aperture, i.e, the near-field intensity at the aperture is improved.

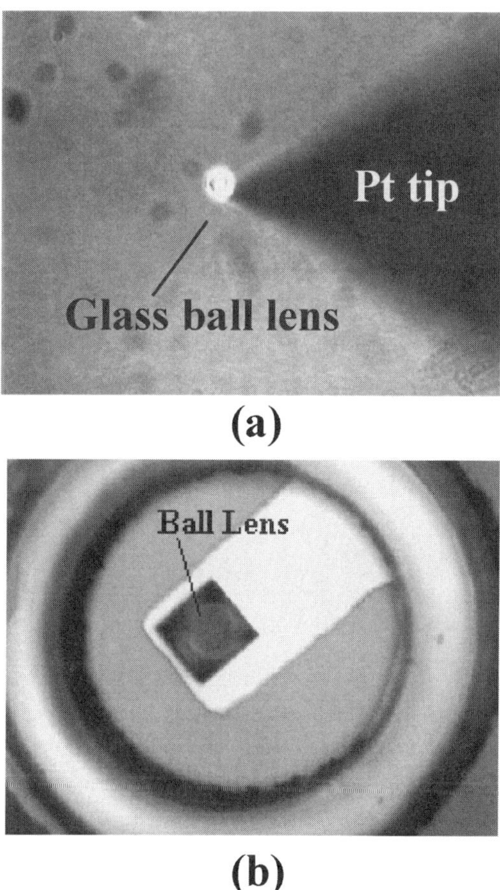

Figure 6.18 (a) Optical image showing the glass ball lens at the end of the Pt tip. The ball is attached to the Pt tip by electrostatic force; (b) optical image of the SiO_2 cantilever on the Si column with a ball lens at the tip.

In case of a hybrid fiber bundle-aperture array (see Figure 6.19), the glass ball lens array can be formed on the upper side of the tip array by spin coating. Details of this method will be explained in Chapter 10. The microlens can be directly formed at the core of the optical fiber bundle by photolithography and heating to form a lens shape and etching, the fast atom beam etching. We have proposed a simple experimental setup for forming a microlens at the core of the fiber bundle as shown in Figure 6.20. One end of the fiber bundle is coated with a negative photoresist by dipping or spray coating. The other end of the fiber bundle is connected with a UV-light source with a connector. The UV light is guided by the core and illuminates the photoresist there. The fiber is next developed in a solvent and heated to about 170°C for 1 hour. After heating, due to the surface tension, the photoresist islands at the cores become softer and photoresist lenses are automatically formed at the cores of the fiber

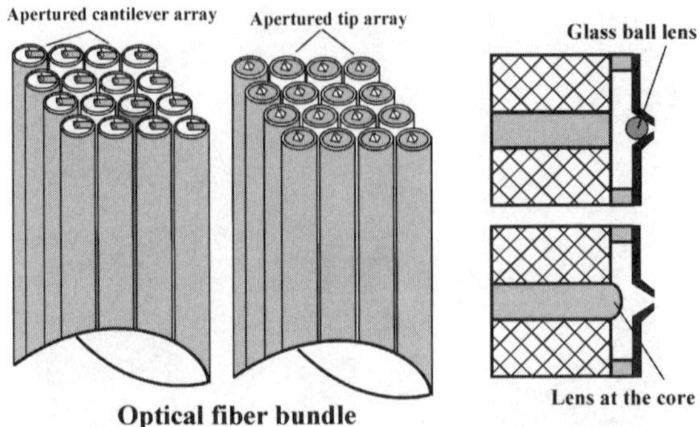

Figure 6.19 Schematic illustration of a hybrid fiber bundle-apertured tip array.

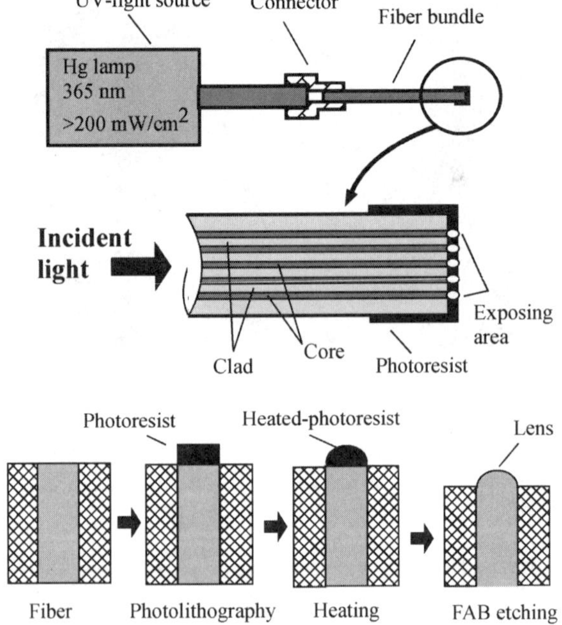

Figure 6.20 Schematic diagram of the fabrication method for forming microlens at the core of the optical fiber bundle by self-photolithography.

bundle. The photoresist at the core can be utilized as a lens. Moreover, the glass lens can be formed at the core of the fiber bundle by dry etching the structure using, for example, a fast atom beam. If the etching rate of the photoresist and SiO_2 of the core are identical, glass lenses at the core can be fabricated. The initial fabricated photoresist lens array at the fiber bundle produced by this technique is shown in Figure 6.21.

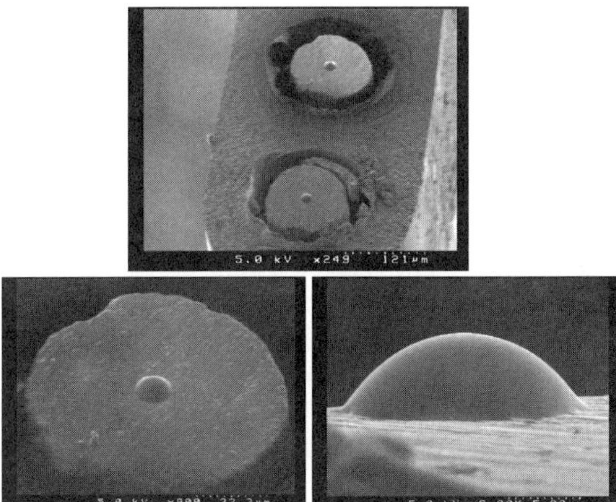

Figure 6.21 Primary results of forming microlens at the core of the fiber bundle using the fabrication process shown in Figure 6.20.

To determine the possibility of using the fabricated photoresist lens for the hybrid fiber bundle-micromachined aperture, the optical transmission through the heated photoresist film was tested. Photoresist film of around 3 µm is coated on the surface of a quartz glass substrate. The sample is next heated at 170°C for 1 hour (the same experimental condition as with the fabrication of the lens). The transmissivity through the sample is next measured with a UV/VIS/NIR spectrometer (JASCO V-570) as shown in Figure 6.22.

It is seen that, for the wavelength that is smaller than 400 nm, because of the high absorption of the photoresist with the light at UV region, the transmissivity strongly depends on the thickness of the photoresist. The

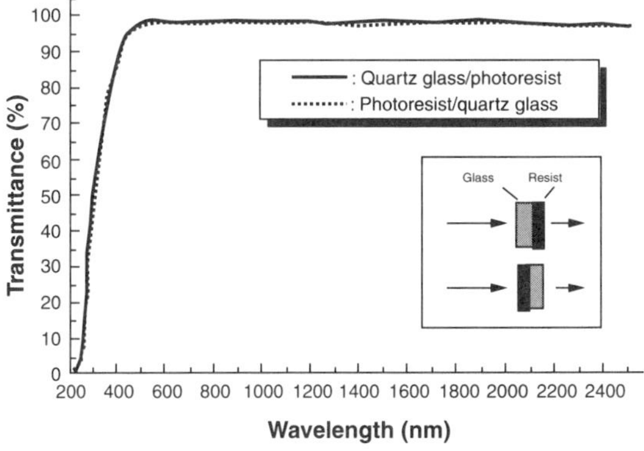

Figure 6.22 Optical transmission through the heated photoresist layer on a quartz substrate.

thicker the photoresist layer, the weaker the transmissivity. However, for the wavelength that is longer than 400 nm, more than 98% of the light is transmitted through the photoresist. The transmissivity is less sensitive to the thickness of the photoresist layer. This means that, for the light with the wavelength that is longer than 400 nm, the heated photoresist lenses at the fiber bundle are applicable. A numerical aperture (NA) of 0.33 and a focal length of 11–12 μm were estimated for the fabricated microlens for the hybrid fiber bundle–micromachined aperture structure.

Using the heated photoresist as a microlens for different applications was presented in the literature.[34-36] The presented technique of forming a microlens at the fiber core is very simple and effective. Lens array can be formed exactly at the core without any alignment. The structure seems applicable not only in hybrid near-field structure, but also for other applications in communication technology with important cost benefits.

Using the fabricated NSOM probe with the cantilever structure at the end of the fiber, optical imaging and other characteristics were demonstrated. Generally, the shear force detection method is utilized for the gap control as well as topography measurements with the fiber probe (see Chapter 2, Section 2.4). With the hybrid structure of the probe, a simpler technique for controlling the tip–sample distance can be utilized. As previously mentioned, because the probe consists of a flexible cantilever at the end, a part of the incident far-field light will be reflected by the tip and some of them return to the core with a certain angle and go back through the fiber. The intensity of the reflected light depends on the position of the cantilever. The bending of the cantilever changes the traveling path and results in a change of the intensity returned to the fiber. The effect of interference between light from the fiber end and that from the tip is possibly negligible because the phase of the light from the tip is diverse. Therefore, the deflection or vibration of the cantilever at the fiber end or the tip–sample distance could be monitored by detecting the reflected light.

To confirm this principle, the following experiment was performed as shown in Figure 6.23. The probe is mounted in a piezo element in a vacuum chamber with pressure of 3×10^{-2} torr. The probe (cantilever) is vertically vibrated by applying a driving voltage to the piezo. The vibration amplitude of the cantilever is monitored by a displacement sensor connected with a network analyzer. A laser diode (multimode, $\lambda = 670$ nm, power: few mW) was coupled with the fiber. The far-field light reflected back by the cantilever tip and passed through the fiber core is detected by a PMT through a photo coupler. The optical signal of the PMT is recorded by the network analyzer. Figure 6.24 shows the mechanical vibration amplitude measured by the displacement sensor (the upper one) and the optical signal detected by the PMT (the lower one). It is shown that the peaks in Figure 6.24 show the mechanical resonant peak of the cantilever. It is seen that a very small mechanical oscillation of 3.5 nm could be detected with this measurement.

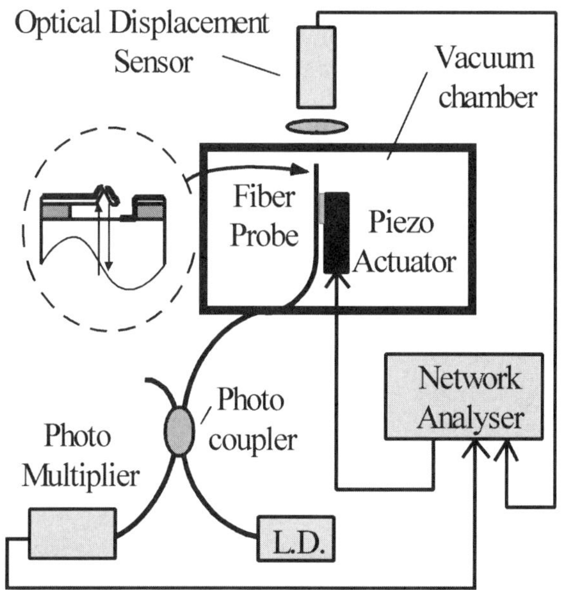

Figure 6.23 Experimental setup for detection of the vibration of the cantilever at the end of the probe.

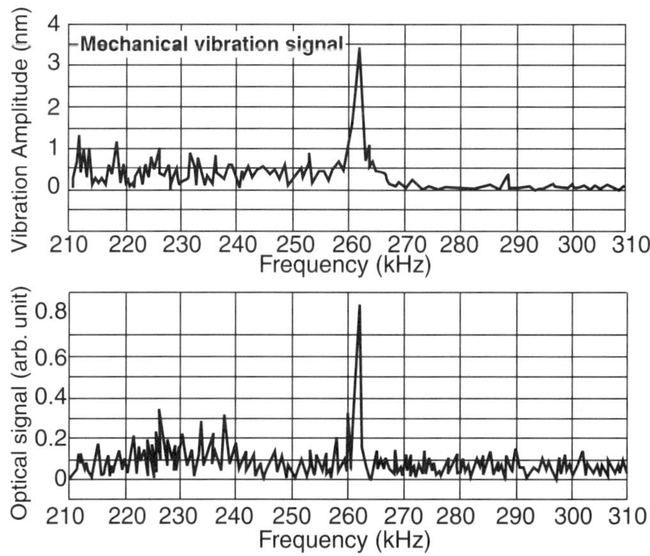

Figure 6.24 Mechanical vibration amplitude of the cantilever recorded by the optical displacement sensor (upper part) and by the reflected far-field light using the PMT (lower part).

Comparing the two plots in Figure 6.24, a detectable minimum deflection of the cantilever of 0.5 nm can be addressed with a reflected far-field signal. This means that the hybrid optical fiber–apertured cantilever can serve as a self-distance modulation probe. By keeping the reflected light intensity at a constant value, one may keep the tip–sample distance constant using a feedback loop. It should be noted, however, that the mechanical quality factor of the cantilever was very low, about 8–10 in atmosphere because of the air damping and the mechanical energy loss of the cantilever. It seems hard to operate in atmosphere. This principle may be useful for NSOM systems in a high vacuum. During fabrication the SiO_2 cantilever was slightly bent due to the internal stress after etching the Si base. However, the bending of the SiO_2 cantilever can be balanced with a metallic opaque layer. During the operation with NSOM, little light will be scattered by the sample and go back through the aperture, however the influence will be small.

Since the probe consists of a cantilever at its end, it can work in contact mode. The instrument of the measurement system is schematically shown in Figure 6.25. The probe with a 300–400 nm aperture is set on a holder and approached until the tip makes contact with the sample that is under observation using an optical microscope. The sample is raster scanned under the tip by applying a driving signal to a XYZ piezo scanner without any Z-feedback control. As mentioned, the flexible cantilever can deflect according to the sample's unevenness. A pulsed laser diode beam ($\lambda = 670$ nm, power: few mW) is coupled into the other end of the fiber probe and illuminates the probe tip to form near-field light at the aperture. Scattered light is collected by an objective, detected by a photomultiplier and the photon signal is lock-in amplified. In this test, a silicon chip on which a resist grid pattern is formed is chosen as a sample. The area of the grid pattern is about 1.4 μm. Figure 6.25b shows the optical near-field image recorded using the described sample, probe and setup. The bright lines in the image correspond to the resist patterns on the silicon. It shows that the NSOM probe can be operated in the illumination mode in which a sample is illuminated by a subwavelength scaled light source with an aperture. Also it is expected that this probe can be utilized in collection mode, in which an evanescent light on a sample is collected by the aperture.

6.5 *Fabrication and characterization of a metallic contact for thermal profiler and thermal recording probe array*

The fabrication process of forming aperture at the apex of the SiO_2 tip is also applicable to a thermal profiler or thermal recording where a nanometric metallic contact at the aperture is utilized. In this section we present the first result of using the aperture structure for application in thermal sensor in nanoscale. The concept of the thermal profilers or thermal recording probe array is shown in Figure 6.26. On the upper side of the apertured SiO_2 tip a Pt/Cr metallic film is formed and an Ni metallic film is formed on the lower

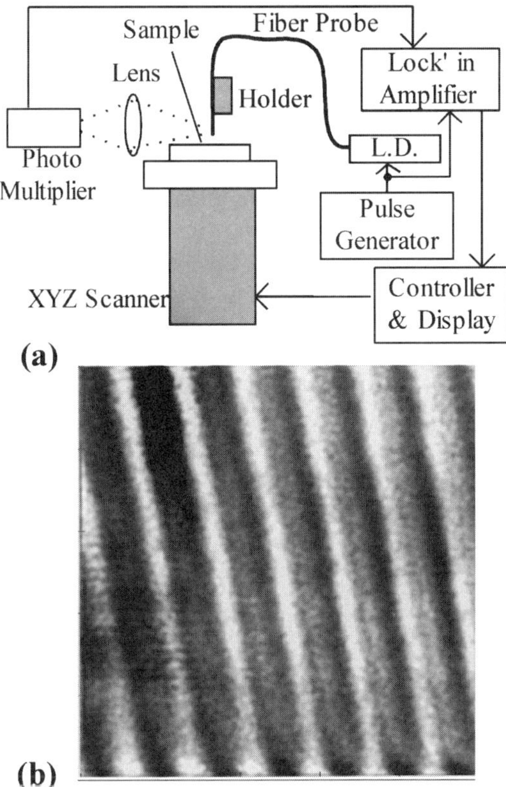

Figure 6.25 (a) Experimental setup for optical near-field imaging in contact mode for the fabricated probe; (b) optical near-field image of the resist patterns on an Si substrate (image size is $5 \times 5 \ \mu m^2$).

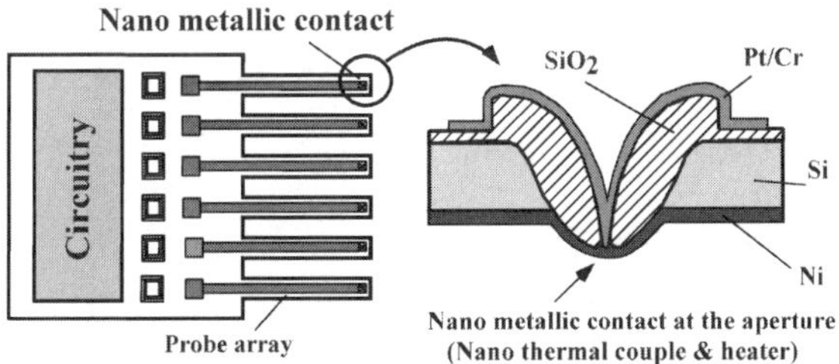

Figure 6.26 Schematic diagram of a nanometric thermal couple or nanoheater integrated probe array for thermal recording.

side. The Pt/Cr and Ni films are isolated from each other by the SiO_2 tip except only at the fabricated aperture. Cr was chosen as an intermediate layer because of its good adhesive properties with the SiO_2. With this structure, the nanometallic contact can serve as a nanosized thermal couple for the nanothermal profiler. Since the electrical resistance is highest at the nanocontact, compared to other regions, by flowing a current, the nanocontact will serve as a nano-heater. Such a probe has been fabricated and details of the fabrication process were reported in reference 37. SEM images of the initial probe with the nano-metallic contact formed at the 150 nm aperture are shown in Figure 6.27. The total thermoelectromotive force of the fabricated probe was measured with changing temperature (Figure 6.28). For a comparison, the thermoelectromotive force of the macroscopic Ni/Pt wire was also simultaneously measured as shown in the figure. It is seen that the thermoelectromotive force of the nanocontact Pt/Ni is identical to the bulk. It should be emphasized that, since the volume of the heater is very small, the thermal response should be very high compared to a macroscopic heater. This should be a very important advantage of the structure, not only for a high-speed, high-resolution thermal profiler, but also for high density data storage where the nanocontact is utilized as a nanosized thermal source for writing, reading and erasing data on, for

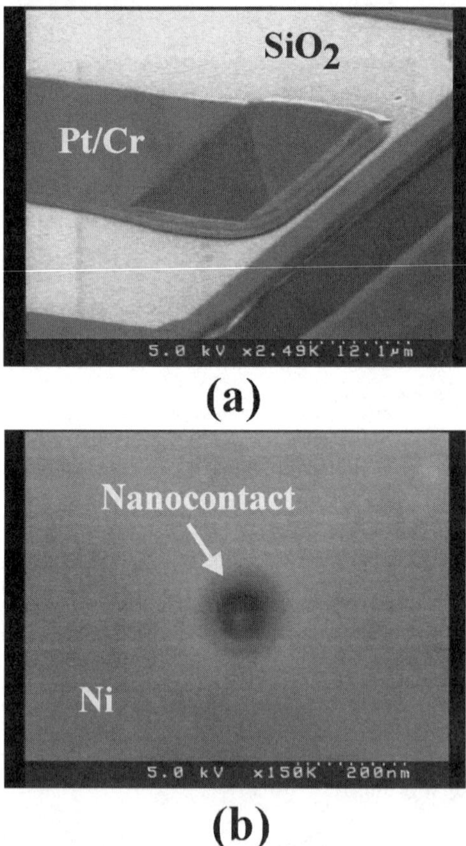

Figure 6.27 SEM images of a fabricated nanometric metallic contact at the aperture of the SiO$_2$ tip (a) from upper side (b) from lower side.

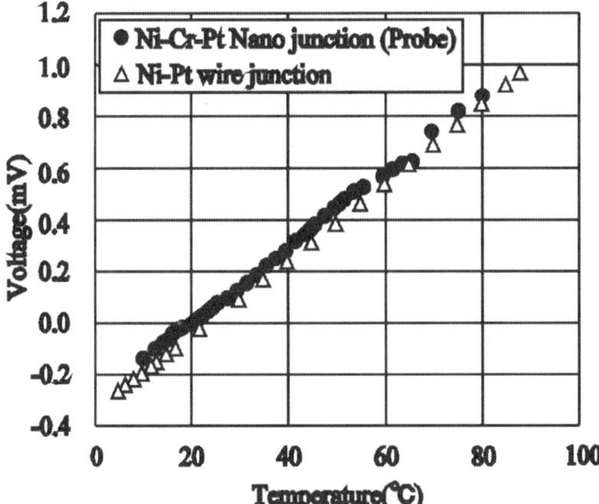

Figure 6.28 Thermo-electromotive force of the fabricated nanosized Ni-Cr/Pt on the probe and the metal wire junction formed from Ni-Pt wire plotted as a function of temperature.

example, phase change material. With the presented concept, a thermal probe array with approximately 30 nm sized heater has been successfully fabricated and an initial result of writing and reading bits on the phase change material has been demonstrated. Temperature of the heater predicted from a change in resistance was 800°C at the flowing current of 4 mA.[38-40]

By flowing a current of about 1 mA through the fabricated metallic contact, the nanoheater is activated and a thermophoton emitted from the tip was observed by a photon-counting camera as shown in Figure 6.29. The thermal response of several microseconds was confirmed. However, to realize the structure, the isolation of the thermal heat from the nanocontact to the other part of the probe should be taken into account.

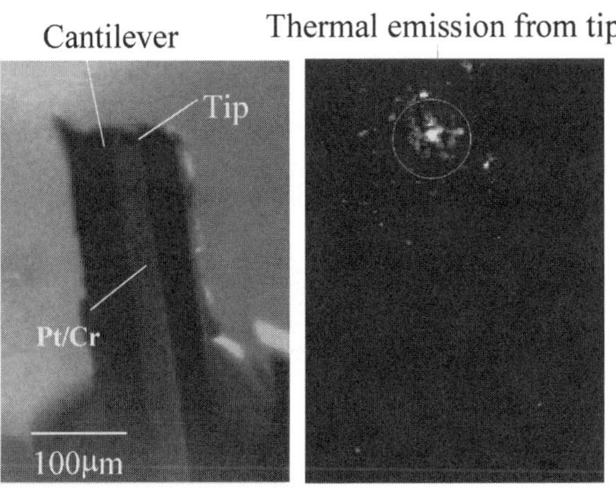

Figure 6.29 Thermophoton emitted from the nanometallic contact at the tip when flowing a current of 1 mA.

6.6 Initial results of the fabrication of electron field emission devices

Field emission devices (FEDs), in which an electron beam is extracted from an emitter under a high electric field, are very important in providing an electron source for applications in flat panel display, multi electron beam lithography, micro-SEM, and electron beam-based analysis devices. The Si micromachining technology has been recognized as an appropriate technique for fabrication of the FED with compactness and low cost, as well as nanoscaled dimensional controls.[41,42] The compactness of such a system would suggest that it be used in an array form to improve throughput and create new applications with electron beams. Recently, new FEDs with carbon nanotubes (CNTs) or doped diamond emitters have gotten much attention due to their brightness and coherent electron sources at a low electric field. In this short section, the updated contributions in the field of fabrication of the FED are not reviewed, but one approach for fabrication of the FED with an integrated electrostatic lens for focusing electron beam is presented. The fabrication technique is based on the presented techniques of forming aperture and selective CNT growing.

The structure of the proposed FED which has an array of a field emitter with an integrated electrostatic lens is shown in Figure 6.30. It consists of sharp Mo tips or CNTs, gate hole array formed on an active Si layer of an SOI (silicon on insulator) wafer and cylindrical electrostatic lens array formed on the bulk Si substrate of the SOI wafer, (see Figure 6.30a). The emitter, gate and lens array are isolated from each other by thermal SiO_2 layers that are 2 μm thick. Since the gate hole can be optimized as small as sub-μm or even sub-100 nm in diameter, a high brightness electron beam is expected to emit from the emitters at a low gate voltage. The electron beam is accelerated and focused by applying sufficient voltage at the lens. Since the lens is made of Si, it is possible to fabricate several lens array layers by the anodic bonding of two SOI wafers with a glass isolated layer as shown in Figure 6.30b. This structure is expected to serve as a compact electron source device for multi electron beam lithography, or other EB-based analysis systems. The system is limited by high voltage up to several kV due to the isolation layer's thickness. However, low-voltage operation can work well in some cases because it can eliminate the proximity effect. Moreover, since the device is very compact its major mechanical components are reduced leading to important cost benefits.

The fabrication process is shown in Figure 6.31. First, the SOI ($Si/SiO_2/Si$) wafer of (7/2/250 μm) and double-sided polishing (step 1) is selected. The wafer is next thermally oxidized for about 100 nm in thickness (step 2). A pyramidal etch pit array is next formed on the active Si layer by wet etching in the TMAH (step 3). The size of the pyramidal etch pit (9 μm) was optimized so that a distance between the top of the etch pit and the SiO_2 layer is about 0.5 μm (see step 3). Next, the SiO_2 layer is removed and the wafer is thermally oxidized in wet oxygen to 2 μm in thickness (1150°C, 6 hours) (step 4). Next, Fe (10 nm thick)/Mo (700 nm thick) patterns are formed on the upper side of

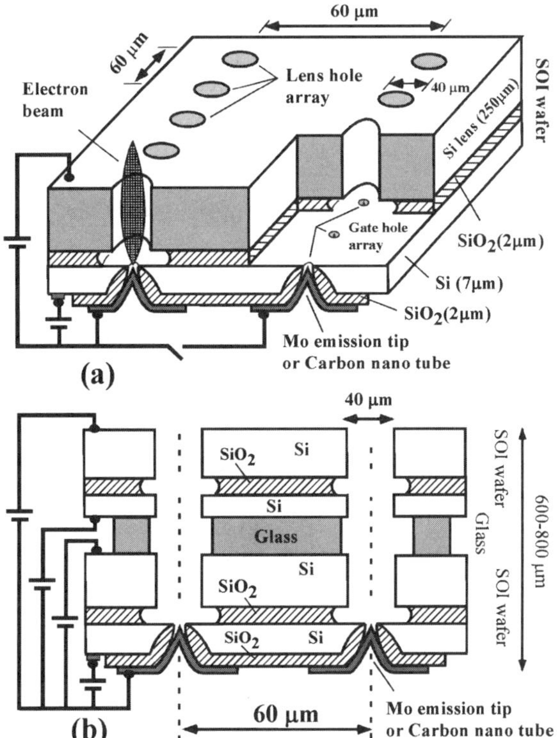

Figure 6.30 (a) Schematic diagram of the electron field emission device (FED) with integrated electrostatic lens formed on the SOI wafer; (b) an extension of the FED structure with multi lenses by anodic bonding two SOI wafers with glass.

the etch pits by photolithography, sputtering and lift-off techniques (step 5). Mo electrical contact pads are also formed at the same time as this step. Here a thin Fe layer for growing CNT is utilized. However, this can be optional using materials, such as W, Mo, SiC and diamond as emittters, as well as other materials as desired by the users. After finishing the structures in the active Si layer, photolithography and dry etching with the ICP-RIE creates lens structures at the substrate side of the SOI wafer. An overetching was done to affirm the formation of lens hole array through the Si substrate (step 6). The wafer is next put in methanol and then in the BHF for etching the SiO₂ until forming a gate hole on the active Si layer (step 7). Finally, the CNT tip can be formed at the apex of the Mo/Fe tip by thermal CVD growing (step 8).

Using the described procedure, the initial FED with integrated electrostatic lens was fabricated. Figures 6.32a and 6.32b show SEM images of the fabricated structure from the front and back, respectively. A close-up of the Mo emitter tip is also indicated in Figure 6.32b. By partly etching the Si lens, the gate hole array is visible as shown in Figure 6.33. The size of the gate hole can be controlled by optimizing the depth of the etch pit formed at step 2 and the thickness of the SiO₂ at step 4. To fabricate gate hole array with

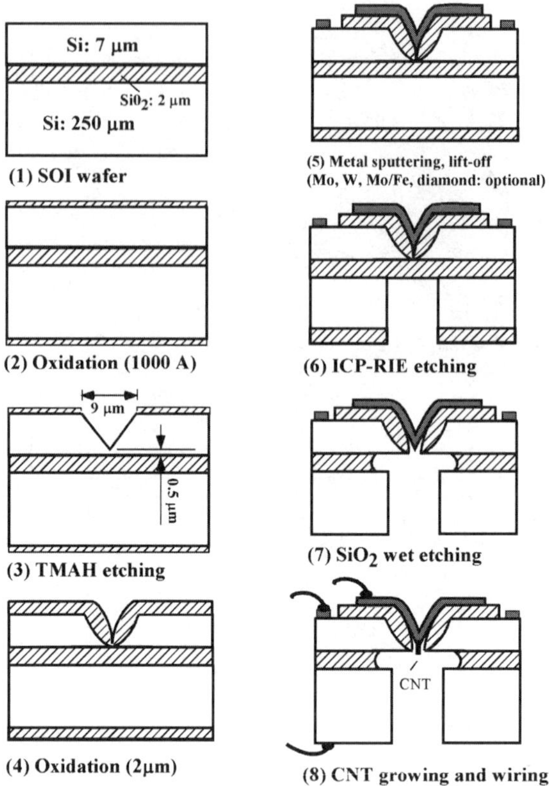

Figure 6.31 Fabrication process of the electron field emission device with integrated electrostatic lens.

high uniformity and submicron diameter, the patterning window for forming the etch pit should use high-resolution electron beam lithography. The emitter array is formed exactly at the center of the gate hole without any alignment.

6.7 Discussion

Enhancement of the near-field intensity is an essential matter in near-field optics. Many attempts have been made to look for a novel structure for near-field applications utilizing coaxial, antenna and metal particle with plasmon excitation. However, to date, no reliable structure has been achieved due to the difficulty of the fabrication. This chapter introduced several fabrication techniques that were developed from the LOSE fabrication process. The final results of such advanced probes have not yet been achieved. However, the possibility of fabrication was confirmed. Furthermore, the fabrication techniques can be extended to metallic contacts with nanometric size for thermal imaging or thermal recording applications. A hybrid structure of an optical

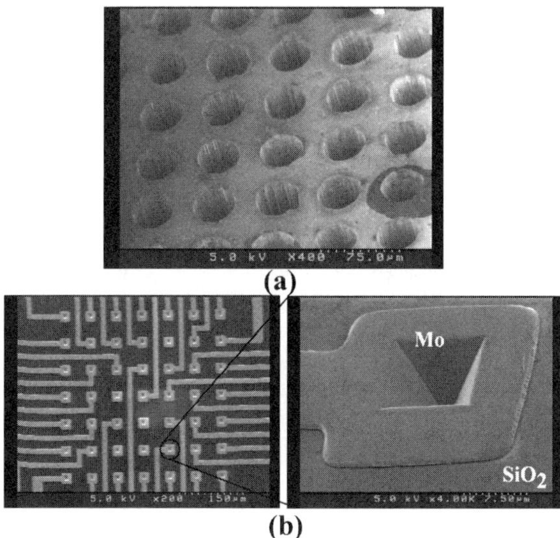

Figure 6.32 SEM images of the fabricated FED: (a) lens hole array; and (b) Mo emitter array and its close-up view.

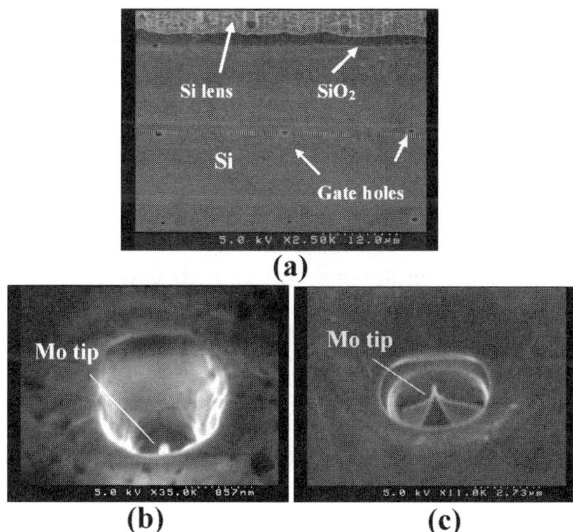

Figure 6.33 SEM images of the fabricated FED: (a) gate hole array and Si lens; (b) submicron gate hole with Mo emission tip; (c) larger gate hole with Mo emission tip.

fiber and apertured cantilever and its array can be fabricated. Since the aperture is a physical aperture, many other applications in nanoscale using photon, ion, electron field emission devices, nano-Raman, or nano-Auger, can also be developed.

References

1. Keilmann, F., U.S. Patent 4, Scanning tip for optical radiation, 994, 818, 1991.
2. Fee, M., Chu, S., and Hansch, T.W., Scanning electromagnetic transmission line microscope with subwavelength resolution, *Op. Comm.*, 69, 219, 1989.
3. Fischer U.Ch. and Zapletal, M., The concept of a coaxial tip as a probe for scanning near-field optical microscopy and steps towards a realization, *Ultramicro.*, 42–44, 393, 1992.
4. Matsumoto, T. et al., Fabrication of a near-field optical fiber probe with a nanometric metallized protrusion, *Opt. Rev.*, 5, 369, 1998.
5. Yatsui, T., Kourogy, M., and Ohtsu, M., Highly efficient excitation of optical near-field on an apertured fiber probe with an asymmetric structure, *Appl. Phys. Lett.*, 71, 1756, 1997.
6. Yatsui, T., Kourogy, M., and Ohtsu, M., Increasing throughput of a near-field optical fiber probe over 1000 times by the use of a triple-tapered structure, *Appl. Phys. Lett.*, 73, 2090, 1998.
7. Bae, J. et al., Experimental demonstration for scanning near-field optical microscopy using a metal micro-slit probe at millimeter wavelengths, *Appl. Phys. Lett.*, 71, 3581, 1997.
8. Oesterschulze, E., On the development and potential of cantilever-based probes for SNOM applications, *Opt. Mem. Neu. Net.*, 7, 251, 1998.
9. Heinz Raether, *Surface Plasmons on Smooth and Rough Surfaces and on Grating*, Springer-Verlag, Berlin, 1988
10. Iijima, S., Helical microtubules of graphite carbon, *Nature*, 56, 354, 1991.
11. Tan, S.J. et al., Room temperature transistor based on a single carbon nanotube, *Nature*, 393, 49, 1998.
12. Stevens, R.M.D. et al., Carbon nanotubes as probes for atomic force microscopy, *Nanotech.*, 11, 1, 2000.
13. Dai, H. et al., Nanotubes as nanoprobes in scanning probe microscopy, *Nature*, 384, 147, 1996.
14. Arie, T. et al., Carbon-nanotube probe equipped magnetic force microscope, *J. Vac. Sci. Technol.*, B18, 104, 2000.
15. Cooper, E.B. et al., Terabit-per-square-inch data storage with the atomic force microscope, *Appl. Phys. Lett.*, 75, 3566, 1999.
16. Choi, W.B. et al., Fully sealed, high-brightness carbon-nanotube field-emission display, *Appl. Phys. Lett.*, 75, 3129, 1999.
17. Ajayan, P.M. et al., Growth morphologies during cobalt-catalyzed single-shell carbon nanotube synthesis, *Chem. Phys. Lett.*, 215, 509, 1993.
18. Thess, A. et al., Crystalline ropes of metallic carbon nanotubes, *Science*, 273, 483, 1996.
19. Li, W.Z. et al., Large-scale synthesis of aligned carbon nanotubes, *Science*, 274, 1701, 1996.
20. Murakami, H. et al., Field emission from well-aligned, patterned, carbon nanotube emitters, *Appl. Phys. Lett.*, 76, 1776, 2000.
21. Miyashita, H. et al., Selective growth of carbon nanotubes for nanoelectro mechanical devices, *Tech. Dig. of 14th IEEE International Micro Electro Mechanical Systems* (MEMS-01), Interlaken, Switzerland, 301, 2001.
22. Spalli, O. et al., Improved tip performance for scanning near-field optical microscopy by the attachment of a single gold nanoparticle, *Appl. Phys. Lett.*, 76, 2134, 2000.

23. Okamoto, T. and Yamaguchi, I., Near-field scanning optical microscope using a gold particle, *Jpn. J. Appl. Phys.*, 36, L166, 1997.

24. Tominaga, J. et al., An approach for recording and readout beyond the diffraction limit with an Sb thin film, *Appl. Phys. Lett.*, 73, 2078, 1998.

25. Tominaga, J. et al., The characteristics and the potential of super resolution near-field structure, *Jpn. J. Appl. Phys.*, 39, 957, 2000.

26. Fuji, H. et al., A near-field recording and readout technology using a metallic probe in an optical disk, *Jpn. J. Appl. Phys.*, 39, 980, 2000.

27. Noell, W. et al., Micromachined aperture probe tip for multifunctional scanning probe microscopy, *Appl. Phys. Lett.*, 70, 1236, 1997.

28. Kim, B.J. et al., Moulded photoplastic probes for near-field optical applications, *Journal of Microscopy*, 202, 16, 2001.

29. Minh, P.N. et al., Hybrid optical fiber-apertured cantilever near-field probe, *Appl. Phys. Lett.*, Vol. 79, No. 19, 2001, in press.

30. Watanabe, H. et al., Near-field optical cantilevered probe with tiny aperture on optical fiber, *Tech. Dig. of the 18th Sensor Symposium-Japan*, 309–312, 2001.

31. Minh, P.N. et al., Hybrid optical fiber bundle and apertured cantilever for optical near-field applications, *Proc. Opt. MEMS-2001*, Okinawa, Japan, forth-coming.

32. Katsumata, T. et al., Fiber optic miniature pressure sensor, *Int. Conf. on Optical MEMS*, 141, 1997.

33. Hirata, K. et al., Silicon micromachined fiber-optic accelerometer for downhole seismic measurement, *Trans. Inst. Elect. Eng. Jpn.*, 120-E, 576, 2000.

34. Kufner M. and Kufner S., *Micro-optics and Lithography*, VUB University Press, Brussels, 1997.

35. Sasaki, M. and Hane, K., Direct photolithography on optical fiber, *Proc. Opt. MEMS*, Hawaii, 149, 2000.

36. Roulet, J.C. et al., Microlens systems for fluorescence detection in chemical microsystems, *Opt. Eng.*, 40, 814, 2001.

37. Takimura, N. et al., Heater integrated micro probe for high density data storage, *Tech. Dig. of the 17th Sensor Symposium-Japan*, 423 (2000).

38. Lee, D.W. et al., Fabrication of microprobe array with sub-100 nm nanoheater for nanometric thermal imaging and data storage, *Tech. Dig. of 14th IEEE International MEMS-01*, Interlaken, Switzerland, 204, 2001.

39. Ono, T. et al., Micromachined probe for high density data storage, *Tech. Dig. of the 4th Pacific Rim Conference on Lasers and Electro-Optics*, Makuhari Messe, Chiba, Japan, II-542, 2001.

40. Ono, T., Minh, P.N., Lee, D.W., Esashi, M., Multi-probe aimed for high density data storage, The Review of Laser Engineering, Special Issue on *Information Technology Industry and Laser Technology*, 29, 516, 2001.

41. Rai-Choudhury, P., Ed., *Handbook of Microlithography, Micromachining and Microfabrication, Vol. 1: Microlithography*, SPIE Press, Bellingham, WA, 1997.

42. Rai-Choudhury, P., Ed., *Handbook of Microlithography, Micromachining and Microfabrication, Vol. 2: Micromachining and Microfabrication*, SPIE Press, Bellingham, WA, 1997.

chapter seven

Simulation using the finite difference time domain (FDTD) method

7.1 Introduction

Near-field optics is based on the interaction of an electromagnetic field with nanostructures in all its applications. To understand the properties of near-field optics devices, the electric field in the near-field region and its interaction with the object have to be calculated. This problem is particularly difficult due to the low symmetry of the near-field geometry and the small size of the system compared to the wavelength. The geometrical optics or the scalar physical optics approximations are no longer valid. Moreover, the near-field light, after interacting with the sample, will be diffracted and converted into far-field light; therefore the calculation becomes more complicated. Near-field simulation means to calculate numerical solutions of Maxwell equations.

A considerable effort has been made to model the behavior of the near-zone fields produced by radiation behind a small aperture[1-4] based on the diffraction theory at small holes developed by Bethe[5] and Bouwkamp.[6] All of these theories have been based on the assumption that the energy is coupled through an aperture in a plane screen. Recently, theoretical calculations for near-field problems that use models closer to the actual geometry of the NSOM probe have been developed. For example, Robert[7] has theoretically proved that the near-field power at the aperture of the conical probe drops off as an R^6 relationship (R is radius of the aperture). The simulations are mostly based on Green's function or multiple multipole (MMP) methods.[8]

Presently, the finite difference time domain (FDTD) is rapidly becoming one of the most widely used computational methods in electromagnetics[9-11] and in near-field optics[12-13] due to the power of the technique. The FDTD is capable of computing an arbitrary three-dimensional geometry that is extremely difficult to analyze by other methods. The FDTD is not new for calculation of microwaves but rather new for near-field optics.

In this work, optical near-field distribution has been calculated at the aperture and optical throughput using the FDTD simulation. The simulation model is built from the actual geometry of the tip.

7.2 *FDTD modeling for optical near-field simulation*

In general, solving Maxwell equations for a microscopic object or probe system is complicated; therefore a model based on experimental result plays an important role. To solve the Maxwell equations numerically, Yee[9] proposed an elegant algorithm as shown in Figure 7.1. The simulation area is divided into many cells with cell size being much smaller than the wavelength of the incident light. Figure 7.1a illustrates how the three-dimensional simulation area of the micromachined NSOM tip is immersed in the FDTD lattices. Figure 7.1b illustrates the positions of the electric and magnetic components in a unit cell $(\delta x, \delta y, \delta z)$ of the FDTD lattice in Cartesian coordinates. The Yee algorithm centers the electric field (E) and magnetic field (H) components in three-dimensional space so that every E component is surrounded by four circulating H components, and every H component is surrounded by four circulating E components in one unit cell as shown in Figure 7.1b.

The computing process is easily understood with the help of the space-time chart shown in Figure 7.2. A leapfrog arrangement where all the E components in the space of interest for a particular time step are calculated and stored in memory using the computed H data previously stored in the computer memory. Then, all the H components in the model space are computed and stored in memory using the E data just computed. The process is repeated with the next time steps until a stability of computation is reached.

Since the E and H components are related together, they cannot be calculated at the same time step. A good way to solve this problem is solving E components at every time step of $m\delta t$ and solving the corresponding H components at every time step of $(m + 1/2)\delta t$ with very small δt (m is an integer). The simulation model is built by assigning the desired values of electrical permittivity (ε) and conductivity to each electric field component of the lattice. Correspondingly, desired values of magnetic permeability and equivalent conductivity are assigned to each magnetic field component of the lattice.

To calculate every E and H component in every cell of the lattice, Yee treats the Maxwell equations as follows. In a linear medium the Maxwell equations are written as:

$$\nabla \times E = -\partial B / \partial t \tag{7.1}$$

$$\nabla \times H = \partial D / \partial t + J \tag{7.2}$$

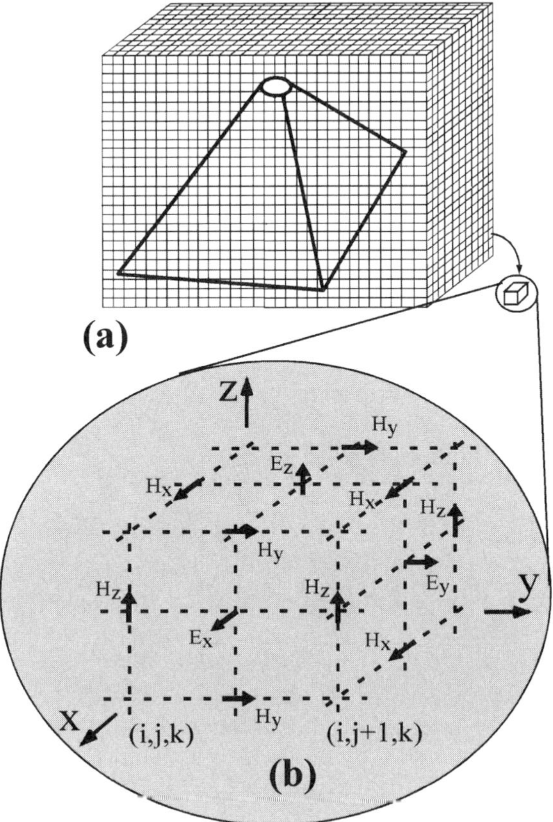

Figure 7.1 (a) Three-dimensional simulation area of the tip embedded in FDTD lattices; (b) Yee cells and positions of electric and magnetic components in one unit cell.

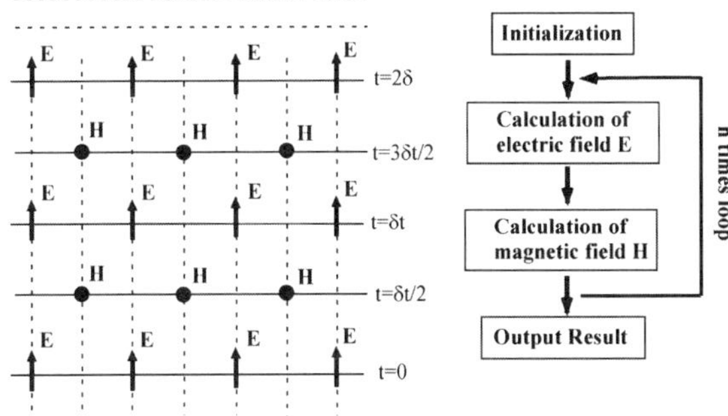

Figure 7.2 Space-time chart of the Yee algorithm.

$$\nabla.D = \rho \tag{7.3}$$

$$\nabla.B = 0 \tag{7.4}$$

Where

$$D = \varepsilon E \tag{7.5}$$

$$B = \mu H \tag{7.6}$$

$$J = \sigma E \tag{7.7}$$

These equations can be rewritten as:

$$\partial E/\partial t = 1/\varepsilon(\nabla x H - \sigma E) \tag{7.8}$$

$$\partial E_x/\partial t = 1/\varepsilon(\partial H_z/\partial y - \partial H_y/\partial z - \sigma E_x) \tag{7.9}$$

$$\partial E_y/\partial t = 1/\varepsilon(\partial H_x/\partial z - \partial H_z/\partial x - \sigma E_y) \tag{7.10}$$

$$\partial E_z/\partial t = 1/\varepsilon(\partial H_y/\partial x - \partial H_x/\partial y - \sigma E_z) \tag{7.11}$$

or,

$$\partial H/\partial t = -1/\mu(\nabla \times E) \tag{7.12}$$

$$\partial H_x/\partial t = 1/\mu(\partial E_y/\partial z - \partial E_z/\partial y) \tag{7.13}$$

$$\partial H_y/\partial t = 1/\mu(\partial E_z/\partial x - \partial E_x/\partial z) \tag{7.14}$$

$$\partial H_z/\partial t = 1/\mu(\partial E_x/\partial y - \partial E_y/\partial x) \tag{7.15}$$

For the sake of simplicity, the electric field E and magnetic field H vectors at a certain space and time to be calculated are defined as:

$$E(x, y, z, t) = E(i\delta x, j\delta y, k\delta z, m\delta t) = E^m(i, j, k) \tag{7.16}$$

$$H(x, y, z, t) = H(i\delta x, j\delta y, k\delta z, m\delta t) = H^m(i, j, k) \tag{7.17}$$

Since the cell size (δx, δy, δz) and the time step size (δt) is very small, so we can approximately define as:

$$\partial F/\partial t \approx [F^{m+1/2}(i, j, k) - F^{m-1/2}(i, j, k)]/\delta t \tag{7.18}$$

$$\partial F/\partial x \approx [F^m(i + 1/2, j, k) - F^m(i - 1/2, j, k)]/\delta x \qquad (7.19)$$

$$\partial F/\partial y \approx [F^m(i, j + 1/2, k) - F^m(i, j - 1/2, k)]/\delta y \qquad (7.20)$$

$$\partial F/\partial z \approx [F^m(i, j, k + 1/2) - F^m(i, j, k - 1/2)]/\delta z \qquad (7.21)$$

where $F = F(x, y, z, t)$ is a general function. $F(x, y, z, t)$ function can be $E(x, y, z, t)$ or $H(x, y, z, t)$.

By substituting the definitions in equations 7.16 to 7.21 into equations 7.9 to 7.11 and equations 7.13 to 7.15, for example, for equation 7.9, we receive:

$$[E_x^{m+1} (i, j, k) - E_x^m (i, j, k)]/\delta t =$$

$$1/\varepsilon[\partial H_z/\partial y - \partial H_y/\partial z - \sigma E_x^{m+1/2} (i, j, k)] \qquad (7.22)$$

Note that on the right-hand side of equation 7.22, including the $E_x^{m+1/2} (i, j, k)$, is not assumed to be stored in computer memory. This problem can be solved by using the so-called semi-implicit approximation:

$$E_x^{m+1/2} (i, j, k) = [E_x^{m+1} (i, j, k) + E_x^m (i, j, k)]/2 \qquad (7.23)$$

By substituting equation 7.23 into equation 7.22, we receive:

$$E_x^{m+1} (i, j, k) = (1 - \sigma\delta t/2\varepsilon)/(1 + \sigma\delta t/2\varepsilon)E_x^m (i, j, k) +$$

$$(\delta t/\varepsilon)/(1 + \sigma\delta t/2\varepsilon) \{[H_z^{m+1/2} (i, j+1/2, k) - H_z^{m+1/2} (i, j - 1/2, k)]/\delta y -$$

$$[H_y^{m+1/2} (i, j, k + 1/2) - H_y^{m+1/2} (i, j, k - 1/2)]/\delta z\} \qquad (7.24)$$

Operating the same way for other components of E and H vectors, the following expressions for Maxwell equations in three dimensions for non-magnetic materials are:[10]

$$E_y^{m+1} (i, j, k) = (1 - \sigma\delta t/2\varepsilon)/(1 + \sigma\delta t/2\varepsilon)E_y^m (i, j, k) +$$

$$(\delta t/\varepsilon)/(1 + \sigma\delta t/2\varepsilon) \{[H_x^{m+1/2} (i, j, k + 1/2) - H_x^{m+1/2} (i, j, k - 1/2)]/\delta z -$$

$$[H_z^{m+1/2} (i + 1/2, j, k) - H_z^{m+1/2} (i - 1/2, j, k)]/\delta x\} \qquad (7.25)$$

$$E_z^{m+1} (i, j, k) = (1 - \sigma\delta t/2\varepsilon)/(1 + \sigma\delta t/2\varepsilon)E_z^m (i, j, k) +$$

$$(\delta t/\varepsilon)/(1 + \sigma\delta t/2\varepsilon) \{[H_y^{m+1/2} (i + 1/2, j, k) - H_y^{m+1/2} (i - 1/2, j, k)]/\delta x -$$

$$[H_x^{m+1/2} (i, j + 1/2, k) - H_x^{m+1/2} (i, j - 1/2, k)]/\delta y] \qquad (7.26)$$

$$H_x^{m+1/2}(i, j, k) = H_x^{m-1/2}(i, j, k) + \delta t/\mu$$

$$\{[E_y^m(i, j, k + 1/2) - E_y^m(i, j, k - 1/2)]/\delta z -$$

$$[E_z^m(i, j + 1/2, k) - E_z^m(i, j - 1/2, k)]/\delta y\} \qquad (7.27)$$

$$H_y^{m+1/2}(i, j, k) = H_y^{m-1/2}(i, j, k) + \delta t/\mu$$

$$\{[E_z^m(i, j, k + 1/2) - E_z^m(i, j, k - 1/2)]/\delta x -$$

$$[E_x^m(i, j + 1/2, k) - E_x^m(i, j - 1/2, k)]/\delta z\} \qquad (7.28)$$

$$H_z^{m+1/2}(i, j, k) = H_z^{m-1/2}(i, j, k) + \delta t/\mu$$

$$\{[E_x^m(i, j, k + 1/2) - E_x^m(i, j, k - 1/2)]/\delta y -$$

$$[E_y^m(i, j + 1/2, k) - E_y^m(i, j - 1/2, k)]/\delta x\} \qquad (7.29)$$

With the system of equations 7.24 to 7.29, the new values of the electric field or magnetic field components at any lattice point depends only on previous values and the previous value of the components of the other field vector at the adjust points as well as the corresponding parameters of ε, μ and σ. This means that the Maxwell equations can be numerically treated at every cell of the simulation area if the combination of ε, μ and σ is known.

To improve the accuracy of the FDTD simulation, the cell size must be very fine compared to the incident wavelength and size of the object and the time step must be small. Since the FDTD solves the Maxwell equations without any simplifying other than the discrete grid, the method is well suited for optical near-field simulations.[12-14] The electric field and magnetic field vectors on every point of the grids are numerically calculated with the help of a computer with sufficient memory and speed. A sinusoidal driving electric or magnetic field is added to the field values in certain cells to provide the source of the wave being modeled. The fields are calculated cell by cell for the whole grids at each time step; then time is advanced by δt to the next time step and new field values are next calculated throughout the grid and so on. In short, the field values at every cell are updated at every time increment, until a steady state is reached (that is the field is not changing peak to peak amplitude). In the FDTD simulation, the number of cells used to model the system and the number of time steps of calculation requires a quite fast computer with a large memory.

7.3   *Results of the FDTD simulation*

For FDTD simulation, it is essential to build a correct model that is as close to the real object as possible. The model used here is based on the cross-section of the tip formed by the focused ion beam milling (for example, as shown in Chapter 6, Figure 6.4b). A simulation area of $200 \times 200 \times 130$ cells, corresponding to an area of $3.2 \times 3.2 \times 2.08$ μm^3 was determined for the three-dimensional FDTD simulation as shown in Figure 7.3a. The medium consists of air, SiO_2 and 100 nm thick Al. A cell size of 16 nm was used for this simulation, which is much smaller than the wavelength used. A time step size of one five-hundredths of an optical cycle was used that satisfies the Courant stability condition in the three-dimensional model:

$$\text{Time step size} \le \text{Cell size}/c(\sqrt{3})\qquad(7.30)$$

Where c is speed of the light. The steady-state solutions are reached in five optical cycles corresponding to around 2600 time steps. We used an S-polarized plane wave (TM mode), where the electric field vector of the incident light was parallel with the cross-section of the tip, at 442 nm wavelength

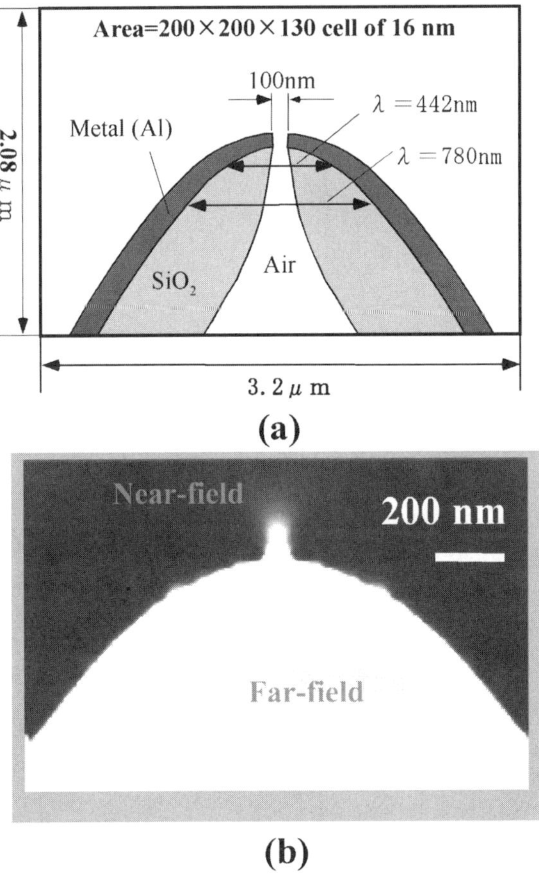

Figure 7.3 (a) Model for FDTD simulation that is built from the SEM image of the FIB cutting; (b) result of FDTD simulation for 96-nm aperture using the described model and conditions.

for our calculation. To deal with a metallic medium (Al opaque layer), a frequency-dependent FDTD method was employed, which recursively solves finite difference.

Maxwell equations using the Drude dispersive model as metallic permittivity. Drude dispersive model for Al permittivity is expressed as:

$$\varepsilon(w) = 1 - w_p^2 / (w^2 + jwg) \tag{7.31}$$

Where wp is plasma frequency: $w_p = 15.565$ eV; g is damping constant: $g = 0.608$ eV. These parameters are taken based on reported results.[15,16] Mur's absorbing boundary condition was used[17] for absorbing the light at the boundary of the simulating region. With the described model (Figure 7.3a) and described conditions, distribution of the optical near-field intensity around a 96 nm diameter aperture is shown in Figure 7.3b.

From the simulated result shown in Figure 7.3b, an optical throughput of approximately 1% for 96 nm aperture was estimated. The estimated optical throughput was defined as the ratio of the near-field and far-field intensities at the center of aperture with a z distance of 10 nm from the aperture. The near-field and far-field intensities were calculated with and without the probe, respectively. The simulated result of optical throughput is in agreement with the evaluation result presented in Chapter 5.

It can be seen that near-field light is very well confined in the aperture area. Figure 7.4 shows an optical profile along the x-axis at the center of 60 nm aperture with various z-values of 0, 10, and 50 nm away from the aperture. A decrease of the near-field intensity at the center of the aperture when increasing the distance in z direction is plotted in Figure 7.5. These results are obtained by simulation with $150 \times 150 \times 120$ cells of 10 nm cell size, corresponding to a simulation area of $1.5 \times 1.5 \times 1.2$ µm³.

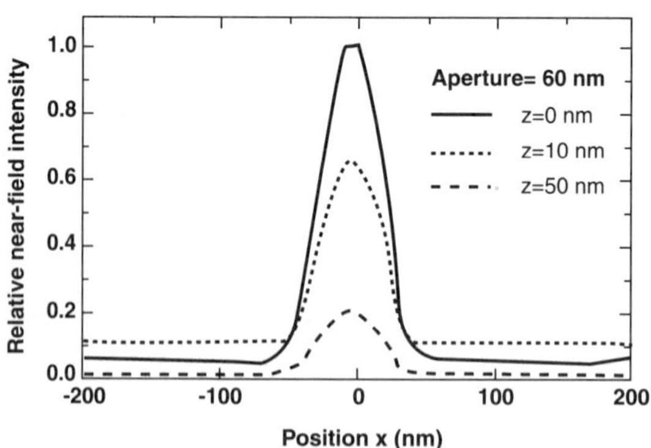

Figure 7.4 Cross section of the near-field intensity at the center of the 60-nm aperture with different z distances.

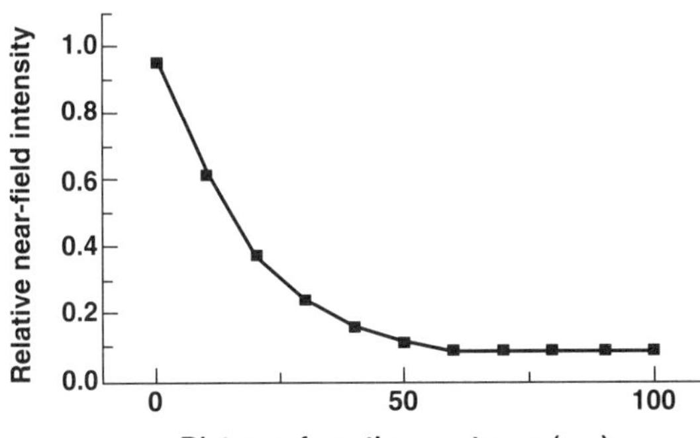

Figure 7.5 Near-field intensity of 60-nm-aperture as a function of z distance.

The simulation results indicated that the optical throughput of the fabricated aperture tip is very high compared to the optical fiber tip. Since the loss is due to the absorption of metal film and the cut-off effect, a simple way to decrease the loss or to increase optical throughput is to shorten the metallic guide or the cut-off region. The fabricated SiO_2 tip yields a high optical throughput because of several reasons as the following:

- Since the metallic opaque layer is formed at the exterior of the tip, the opening angle of the tip is very large. The cut-off effect is minimized or does not exist.
- The aperture is empty (air); therefore incident light in TEM_{00} mode can freely come to the aperture without strong losses in the tip.
- The shape of the SiO_2 tip may serve as a convergent lens to concentrate more light into the aperture.
- Compared to the optical fiber tip, the transmission length inside the SiO_2 tip is very short so the loss is minimized.

As mentioned in Chapter 6, optical throughput can be improved by utilizing the coaxial tip or the aperture with a small metallic scatterer at the center of the aperture. We have also simulated the near-field pattern for coaxial aperture structure. The simulating models for 100 nm aperture and 100 nm coaxial aperture NSOM tips are shown in Figures 7.6a and 7.6b, respectively. The calculated near-field intensities as a function of z distance in both cases are shown in Figure 7.7. Although the enhancing factor is not large (4 times at z larger than 80 nm), improvement of the optical throughput is observed. Although the results presented in this chapter are still preliminary, the FDTD seems very effective for simulation of the near-field problem. Since the shape and the size of the microfabricated NSOM tip and aperture is very well defined, the simulation work with the FDTD method is effective.

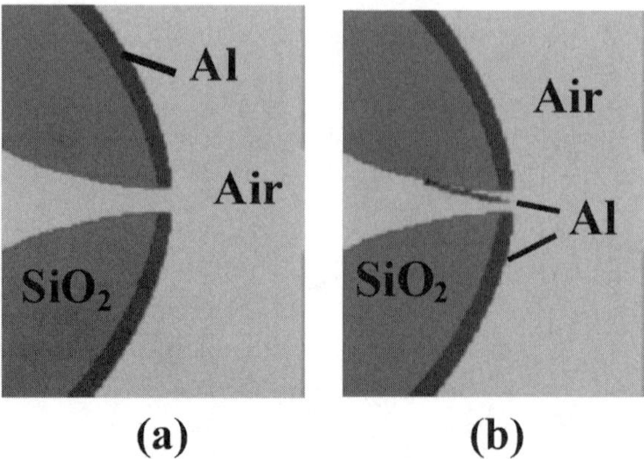

Figure 7.6 (a) FDTD model for 100-nm aperture; (b) FDTD model for 100-nm coaxial aperture.

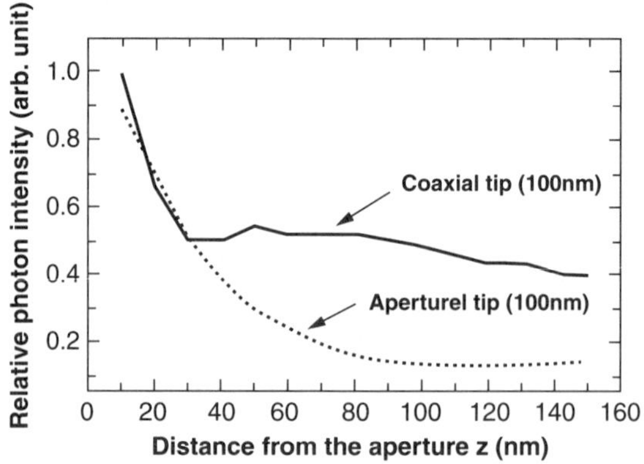

Figure 7.7 Near-field intensity as a function of z-distance away from the aperture in z direction for aperture tip (dotted line) and coaxial tip (solid line).

References

1. Leviatan, Y., Study of near-zone fields of a small aperture, *J. Appl. Phys.*, 60, 1577, 1986.
2. Robert, A., Electromagnetic theory of diffraction by a circular aperture in a thick, perfectly conducting screen, *J. Opt. Soc. Am.*, A4, 1970, 1987.
3. Robert, A., Near-zone fields behind circular apertures in thick, perfectly conducting screens, *J. Appl. Phys.*, 65, 2896, 1989.
4. Robert, A., Small-hole coupling of radiation into a near-field probe, *J. Appl. Phys.*, 70, 4045, 1991.
5. Bethe, H.A., Theory of diffraction by small holes, *Phys. Rev.*, 66, 163, 1944.
6. Bouwkamp, C.J., Theoretical and numerical treatment of diffraction through a circular aperture, *IEEE Trans. Ant. Prop.*, AP-18, 152, 1970.

7.   Grober, R.D. et al., Modal approximation for the electromagnetic field of a near-field optical probe, *Appl. Opt.*, 35, 3488, 1996.

8.   Novotny, L., Pohl, D.W., and Hecht, B., Light confinement in scanning near-field optical microscopy, *Ultramicros.*, 61, 1, 1995.

9.   Yee, K.S., Numerical solution of initial boundary value problems involving Maxwell's equations in isotropic media, *IEEE Trans. Ant. Prop.*, 14, 302, 1996.

10.   Taflove, A., *Computational Electrodynamics: The Finite-Difference Time-Domain*, Artech House, Boston, 1995, p. 65.

11.   Kunz, K.S. and Luebbers, R.J., The finite difference time domain method for electromagnetics, CRC Press, Boca Raton, FL, 1993.

12.   Christensen, D.A., Analysis of near-field tip patterns including object interaction using finite-difference time-domain calculations, *Ultramicros.*, 57, 189, 1995.

13.   Furukawa, H. and Kawata, S., Analysis of image formation in a near-field scanning optical microscope: effects of multiple scattering, *Opt. Comm.*, 123, 170, 1996.

14.   Furukawa, H. and Kawata, S., Local field enhancement with an apertureless near-field microscope probe, *Opt. Comm.*, 148, 221, 1998.

15.   Novotny, L. and Hafner, C., Light propagation in a cylindrical waveguide with a complex, metallic dielectric function, *Phys. Rev. E*, 50, 4094, 1994.

16.   Bian, R.X. et al., Single molecule emission characteristics in near-field microscopy, *Phys. Rev. Lett.*, 75, 4772, 1995.

17.   Mur, G., Absorbing boundary conditions for the finite-difference approximation of the time domain electromagnetic-field equations, *IEEE Trans. Electromagn. Compatib.*, EMC-23, 377, 1981.

chapter eight

Subwavelength optical imaging with fabricated probes

8.1 Introduction

In Chapter 2 two common configurations of near-field imaging, the NSOM and PSTM, were presented. In the NSOM configuration, the incident light is forced on a subwavelength aperture. The near-field light located near the aperture is used for illuminating the sample surface. Depending on the opacity of the sample, the transmitted or reflected photon by the surface is detected by a photodetector. Since the near-field intensity strongly depends on the aperture–sample distance, the aperture is normally kept at a constant distance in proximity to the surface. In the PSTM configuration, a sharp dielectric or metallic tip is typically used to scan in proximity to the sample located on the surface of a prism under the total internal reflection illumination. The tip acts as a scatterer to disturb and convert the near-field light into detectable far-field light.

Since the fabricated probes used here are apertured cantilevers, the NSOM configuration was employed. The fabricated probes are used for AFM/NSOM and capacitive–AFM/NSOM. The fabricated probes are operated with a commercial AFM system (Olympus-NV2000). The aperture–sample distance is kept constant by the AFM technique. The near-field signal is detected by the photomultiplier tube (PMT) and acquired by the AFM controller. With this system it is possible to observe topographic and optical images simultaneously by monitoring the vertical position z of the sample and corresponding optical signal as a function of x, y positions. The system was also modified for other experiments of thermal imaging, near-field photolithography, and near-field recording as well.

This chapter demonstrates the near-field optical imaging of several surfaces using the fabricated probes with different aperture size in reflection, transmission, contact, and dynamic modes.

8.2   Measurement setups

The schematic illustration of the AFM/NSOM measurement setup is shown in Figure 8.1. The fabricated probe is mounted on the AFM stage. A light beam from a laser diode, 780 nm wavelength and approximately 2 mW power, is directly focused on the tip area with an objective lens (50x, 0.55 NA). A part of the incident light is reflected by the tip's walls and monitored by a deflection detector for AFM signal. The near-field light generated at the aperture is reflected by the surface of the sample, collected by a long-distance working lens (20x, 0.28 NA), and detected by a photomultiplier (Hamamatsu-H5783). The signal from the output of the PMT is amplified and acquired by the AFM controller as an optical near-field signal. The sample is located on an XYZ piezoelectric scanner. For observation of the probe in the side view when approaching the surface of the sample, a charge-coupled device (CCD) camera connected to a monitor is used. The sample is horizontally scanned in x and y directions. With this setup the AFM in contact mode and NSOM signal can be observed simultaneously.[1,2]

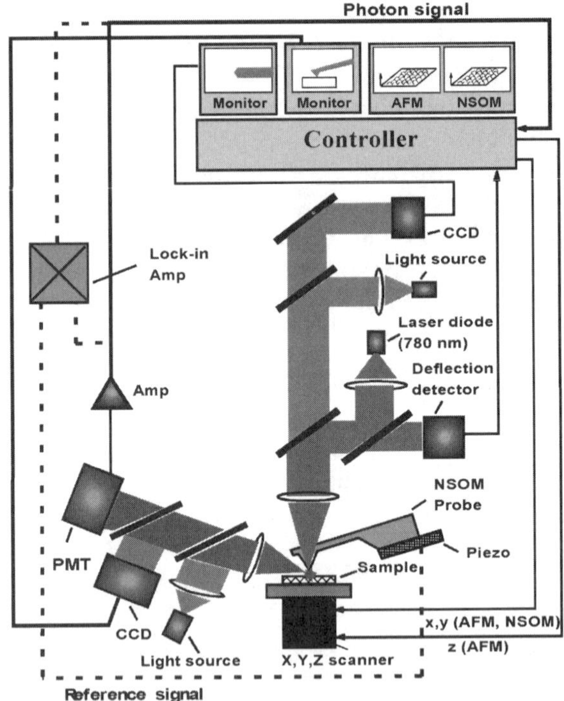

Figure 8.1 Measurement setup of the AFM/NSOM system for the fabricated probe in contact and tapping modes. The setup is based on the commercial AFM system (Olympus-NV2000). The dotted lines in the figure are for tapping mode only.

The setup can also be used for the AFM/NSOM in dynamic mode (see Figure 8.1, dotted lines). The probe is mounted on the noncontact AFM stage on which the cantilever can be vibrated by a piezo element or electrostatic actuation with the capacitive–AFM/NSOM probe. The near-field signal is lock-in amplified to extract the light signal component that is modulated at the same frequency as the probe vibration. The output of the lock-in is transferred to the controller as the optical near-field signal. The tip–sample distance is kept constant by utilizing the AFM tapping mode modulation. The amplitude of vibration is monitored by detecting the deflection of the cantilever with a laser beam reflected from the surface of the cantilever. The near-field image is reconstructed with the photon signal for each x–y position.

The fabricated probe can also be operated in transmission mode. The experimental setup of the AFM/NSOM in transmission mode is schematically shown in Figure 8.2. The probe is mounted on the AFM stage. An He-Cd laser (442 nm wavelength) is focused on the tip area with the same objective lens of the AFM system. AFM laser of 780 nm wavelength is focused on the cantilever for monitoring the deflection of the cantilever for AFM signal. A transparent sample is located on a prism. Both the sample and prism are placed on the XYZ piezoelectric scanner. The near-field light generated from the aperture is collected by a long-distance lens and detected by the photomultiplier tube after transmitting through the sample and the prism. The collected lens is aligned and focused on the prism with the help of a monitor-equipped CCD camera. This setup is also applied for demonstration of the optical recording that will be explained more in the next chapter.

8.3　Measurement results

Using the described measurement setups above, topographic and near-field optical images of several surfaces have been recorded using fabricated probes with different aperture sizes. Figures 8.3 and 8.4 illustrate the AFM and corresponding NSOM images of photoresist and Au patterns on Si surfaces that were observed in contact mode using the aperture of 400 nm diameter.[1] The sample of latex balls on mica substrate for imaging is prepared as follows (see Figure 8.5a). Latex spheres of approximately 300 nm diameters are immersed in DI-water. Next, latex balls contained water drops are dropped on the mica surface having a water absorber in the rear of the sample. Immediately water in the solution is quickly absorbed by the water absorber. Most of the latex spheres move toward the absorber at a high speed. A monolayer of latex spheres at the center of the sample is then obtained. Finally, a thin Cr film of about 10 nm is deposited on the sample. An SEM image of the prepared sample of Cr coated 300 nm latex sphere monolayer is shown in Figure 8.5b.[3]

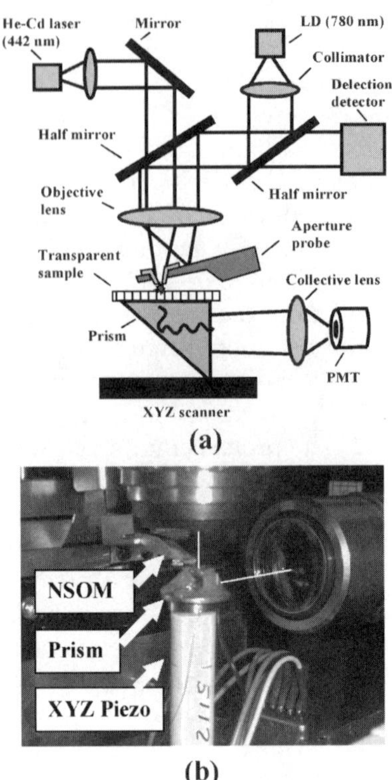

Figure 8.2 (a) Measurement setup of the AFM/NSOM system for the fabricated probe in transmission mode; (b) optical picture of the setup.

Using probes with apertures of approximately 130 nm diameter (determined by SEM), the AFM in contact mode and corresponding NSOM images of Cr coated latex balls on mica substrate were observed as shown in Figure 8.6. AFM and NSOM images of 200 nm Cr grating on glass substrate were simultaneously observed as shown in Figure 8.7.[2] Note from Figure 8.7 that very small Cr particles of about 15 nm (determined by SEM) can be observed with 130 nm aperture. The wavelength of the incident far-field light was 780 nm. This means that a resolution of $\lambda/52$ can be achieved with our fabricated probe and the described measurement setup. The images shown in Figures 8.6 and 8.7 were recorded by scanning at a speed of 2 sec per 1 scanning line. The probes were scanned many times on the sample without damaging either the sample or the tip. This means that the metal coated SiO_2 tip shows sufficient mechanical hardness to reduce tip wear due to friction force between the tip and the sample during scanning in contact mode.

The probe is also measured in noncontact mode. The system is operated in slope detection method with a scan speed of 1 s/line. The cantilever is vibrated at an amplitude of 100 nm. Using the fabricated probe with an aperture size of 150 nm, cantilever length of 400 μm, width of 150 μm,

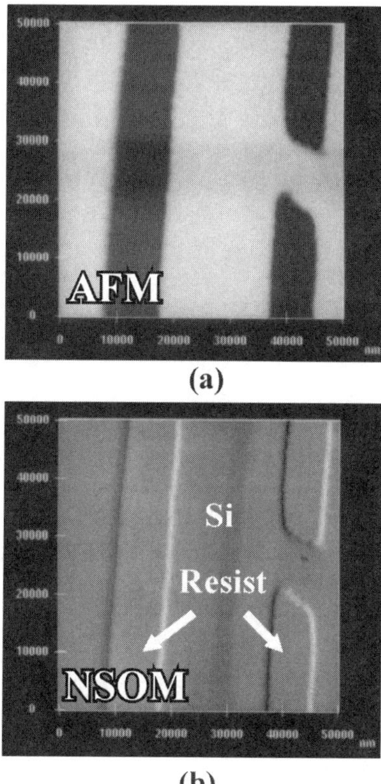

(a)

(b)

Figure 8.3 AFM and corresponding NSOM image of resist patterns on Si surface using 400-nm aperture fabricated probe.

thickness of 4 µm, resonant frequency of 30 kHz and a Q factor of 125, the near-field image of the Cr coated latex balls on mica substrate was recorded as shown in Figure 8.8. The shape of the latex sphere shown in Figure 8.8 is not very round due to the drift of the stage. However, it does offer a possibility of imaging in tapping mode for soft surfaces such as biosamples (e.g., viruses or proteins).[2]

8.4 Discussion

AFM and corresponding NSOM images of several surfaces was demonstrated with the fabricated probes and measurement setups in contact, dynamic, reflection and transmission modes. A resolution of 15 nm ($\lambda/52$) was resolved. In general, the contrast mechanism in near-field imaging is very complicated. The near-field image is sensitive to many parameters including those of the probe, sample and detector. The sample is characterized by its topography and its optical properties (complex index, optical spectrum, Raman spectrum and fluorescent spectrum). The incident field on

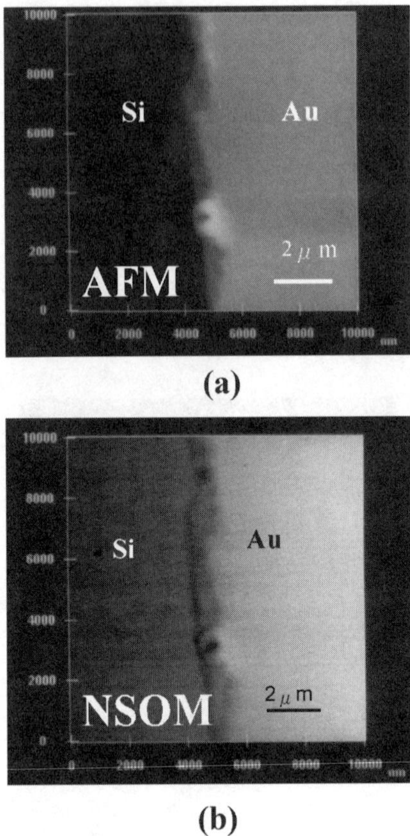

(a)

(b)

Figure 8.4 AFM and corresponding NSOM image of Au pattern on Si surface using 400-nm aperture fabricated probe.

the sample is characterized by the probe, its spatial and temporal coherent, and its polarization and power. The topography may cause a certain effect on the optical information of the surface. A full understanding of such variables requires further study.

It is shown that with fabricated probes, optical imaging with resolution much smaller than the aperture can be achieved. This may be due to the highly localized near-field light at the center of the aperture or to the geometrical structure of the tip. The results of the spatial distribution measurement and the FDTD simulation in previous chapters showed a sharp peak of near-field light at the center of the aperture. That is the reason why a resolution of smaller than the diameter of the aperture can be observed in near-field optical imaging. The presented setup and fabricated probes are expected to be utilized not only for optical imaging, but also for spectroscopy.

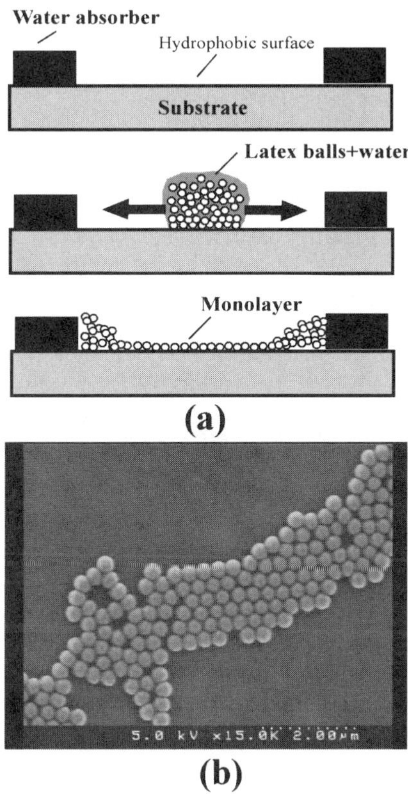

Figure 8.5 (a) Forming a sample of monolayer of latex balls on substrate for NSOM imaging; (b) SEM image of the fabricated latex balls monolayer.

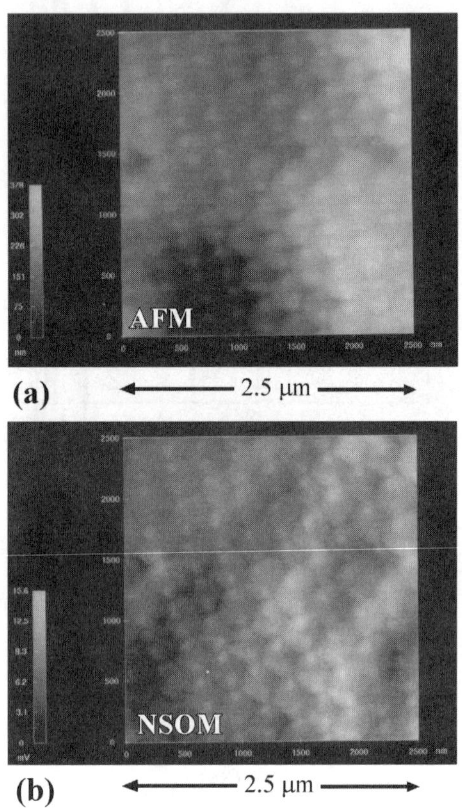

(a)   ◄——— 2.5 µm ———►

(b)   ◄——— 2.5 µm ———►

Figure 8.6 (a) AFM and (b) corresponding NSOM images of Cr coated latex balls on mica surface using 130-nm aperture fabricated probe (scanning area is 2.5×2.5 µm^2).

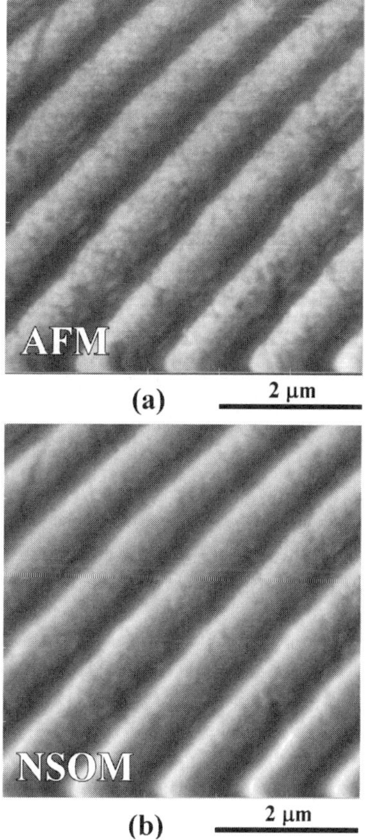

Figure 8.7 (a) AFM and (b) corresponding NSOM images of Cr grating on glass substrate using 130-nm aperture fabricated probe (scanning area is 5×5 μm^2).

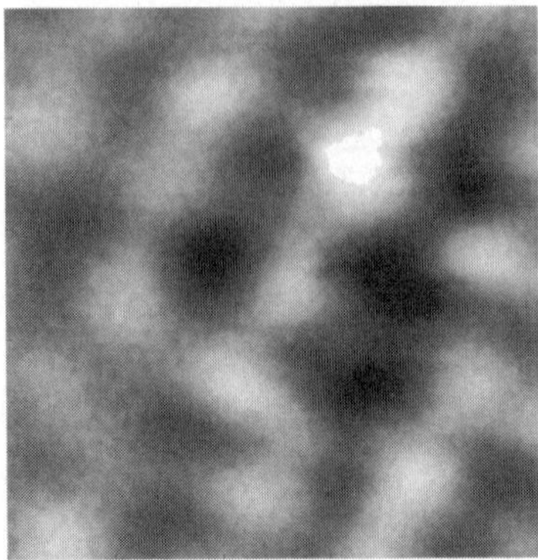

Figure 8.8 NSOM image of Cr coated latex balls on mica surface using 150-nm aperture fabricated probe in taping mode. The white round parts show the latex spheres and some black round holes show the positions of missing latex balls (scanning area is 1.5×1.5 μm^2).

References

1. Minh, P.N., Ono, T., and Esashi, M., Microfabrication of miniature aperture at the apex of SiO$_2$ tip on Si cantilever for near-field scanning optical microscopy, *Sens. Actu.*, A 80, 163, 2000.
2. Minh, P.N., Ono, T., and Esashi, M., High throughput aperture near-field scanning optical microscopy, *Rev. Sci. Instr.*, 71, 3111, 2000.
3. Minh, P.N., Ono, T., and Esashi, M., Nonuniform silicon oxidation and application to the fabrication of apertures for near-field scanning optical microscopy, *Appl. Phys. Lett.*, 75, 4076, 1999.

chapter nine

Optical near-field lithography

9.1 Introduction

The most significant aspect of near-field optical technology is its extreme localization, which provides a powerful scientific tool for observation of small objects in subwavelength or nanoscale size, such as scanning microscopy. Another important aspect of near-field technology is the creation of nanoengineering tools. Much effort has been made to create nanometer scale patterns (lithography)[1-6] and high density data storage systems using near-field light.[7-8]

The rapid development of lithographic techniques followed the requirements for decreasing the minimum size of production devices, especially for semiconductor devices. It has been considered difficult in principle that the resolution of the conventional optical lithography is limited by the far-field effect of diffraction. Therefore, studies on lithography have been performed at shorter wavelength radiations using i-line, extreme ultraviolet (EUV), electron beam, and X-ray radiations. Studies to control the far-field effects have also been made by phase shift mask technique. On the other hand, direct writing methods with high lateral resolution using an electron beam (EB) and a scanning tunneling microscope (STM) have been widely studied, owing to the abilities of high resolution.

As mentioned, near-field light can be created on an extremely localized area, one in which the size is much smaller than the wavelength. Therefore, it is expected that near-field light can be applied to advanced processing beyond the diffraction limits of the light. The advantage of near-field lithography technology is that this technique can be an extension of well-developed modern photolithographic techniques and does not require a shorter wavelength.

Indeed, a near-field scanning optical microscope (NSOM) can be applied to optical lithography. Also utilized is the fabricated apertured probe for writing patterns on a photoresist. However, the technique has the problem of low production throughput. Near-field photolithography with a single NSOM probe seems ineffective unless an array of the probe is utilized. To improve the production throughput many techniques, such as SCALPEL (scattering with angular limitation in projection electron beam lithography),[9,10] nanoimprint lithography,[11] and multibeam lithography,[12] have been studied.

Instead of a single apertured probe, this chapter will describe a technique for pattern transfer beyond the diffraction limit of light using the near-field formed near a small metal aperture and slit.

9.2   Fabrication of nanoscale apertures and slits

Electron beam lithography is used to fabricate narrow metal apertures and slits on a metal coated glass substrate. The near-field is created near the small aperture and narrow slit by illuminating light from the back of the plate. The fabrication procedure is as follows. An electron beam resist (SAL601, Shipley, 300 nm thick) is coated on a glass plate. Then, patterns (apertures or slits) are formed by EB lithography. Before EB exposure, a conductive polymer (Espaser 100, Showa Denko) is spin-coated on the EB resist to prevent buildup charge that causes mis-patterning. Consequently, a thin Cr film (40 nm-thick) is deposited by sputtering onto the patterned EB resist, and lift-off is performed. By narrowing the resist pattern using oxygen plasma etching, a 50 nm-wide slit pattern (minimum size was 15 nm) is fabricated as shown in Figure 9.1.

Another technique uses an atomic force microscope (AFM)-based lithography that is applied to create metal apertures and slits by modifying a thin metal film on a glass substrate. Briefly, a 40 nm-thick metal Cr film is formed on a glass substrate by sputtering. A conductive silicon probe on which Pt is sputtered is used for the AFM-based lithography by applying a voltage between the substrate and the probe as shown in Figure 9.2a. The probe is scanned over the substrate at constant force and scanning speed under an optical microscope. When a voltage of 2V was applied between the substrate and the cantilever with a scanning speed of 20 μm/s, a transparent area (interaction area) is formed. By applying a pulse voltage, many apertures that had a minimum diameter below 50 nm were successfully formed on the Cr film. See the SEM image of Figure 9.2b. This technique can be applied to fabricate not only apertures but also metal slits. By applying DC voltage, a line was formed (see Figure 9.2c). These phenomena are unclear, although it is speculated that Cr atoms are ionized and diffused into the glass by the high electric field at the AFM tip.

9.3   Near-field optical pattern transfer

In near-field scanning optical microscopy a high resolution is achieved by placing a small aperture at a close distance from the surface. By scanning the

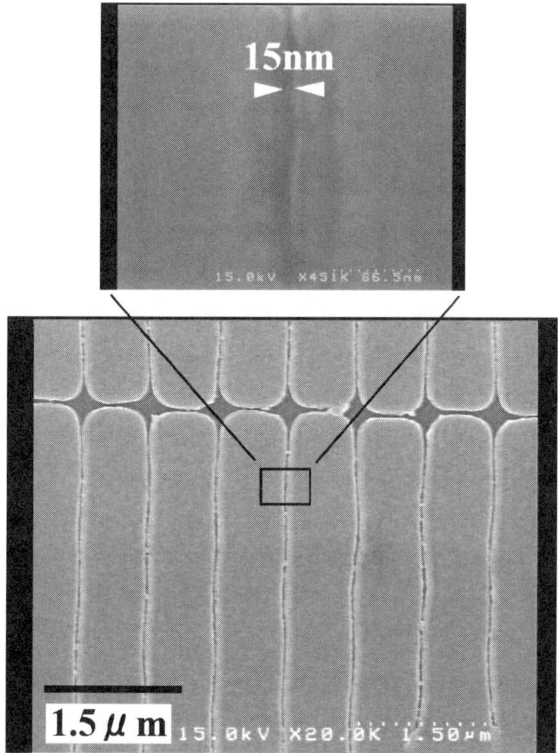

Figure 9.1 SEM image of metal slits fabricated by EB lithography.

aperture it is possible to create patterns lithographically. These experiments are generally performed using an aperture probe. Similarly, if near-field light can be localized on a narrow slit or a surface, it is possible to apply this structure to nanolithography. Near-field lithography is defined here as a technique to transfer patterns using evanescent or near-field light beyond the diffraction limits of light. In this technique the diffraction is not a limiting factor in determining the spatial resolution of optical processing. Localization of near-field light and approach to the sample are considerable factors in near-field lithography.

Pattern transfer with near-field light is demonstrated using a narrow slit or a dot pattern on the mask fabricated by EB lithography. A chemically amplified photoresist (positive type, TSMR V90, Tokyo Ohka Kogyo, Co.) 1 μm in thickness is spin-coated on a silicon substrate. Then, the mask is brought into contact with the substrate and exposed to UV light using a mercury lamp of which main peak is 436 nm (G-line). After development, the resist patterns are observed by AFM. Figure 9.3 shows the AFM image of the resist pattern after lithography. The width and pitch of the grid of the parallel slit on the mask are about 100 nm and 800 nm, respectively. The width of the transferred pattern is about 170 nm at a minimum and the depth is about 200 nm. Here,

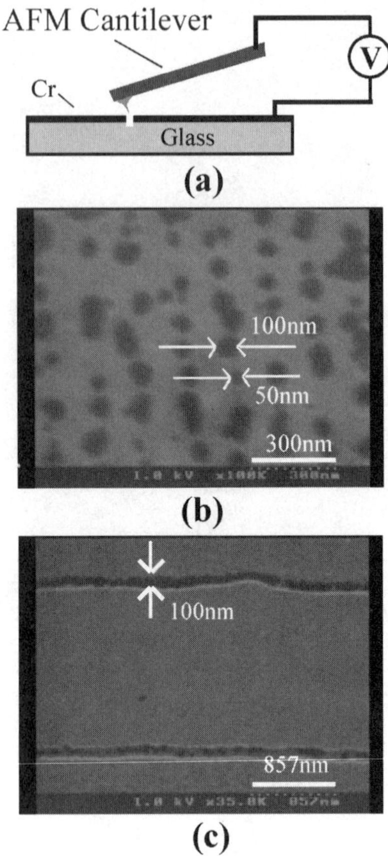

Figure 9.2 Schematic diagram of (a) AFM-based lithography to fabricate nanoscale apertures; (b) SEM image of fabricated apertures; and (c) slits.

the intensity of the mercury lamp is about 1 mW and exposure time is 3 min. Post-exposure baking was done at 105°C for 1 min. Figure 9.4 shows the SEM images of another transferred grid and dot pattern. The exposure on a 300-nm-thick resist was performed under the same exposure conditions as the above experiment. It was found that the minimum sizes were below 150 nm from Figures 9.4a and 9.4b. The diameter of dot aperture and the width of the slit on the metal mask were about 100 nm in this case. The resist pattern appears bright on the silicon background in both figures. The transferred pattern is somewhat larger than the mask pattern because near-field distribution spreads over the aperture. The width characterized by AFM was in good agreement with that measured by SEM. Figure 9.4c shows the cross-section of a 500 nm pitch grid pattern on the resist that was transferred under the same condition as described above. It was found that the width of the exposed lines was about 200 nm and the depth about 350 nm. The obtained pattern width

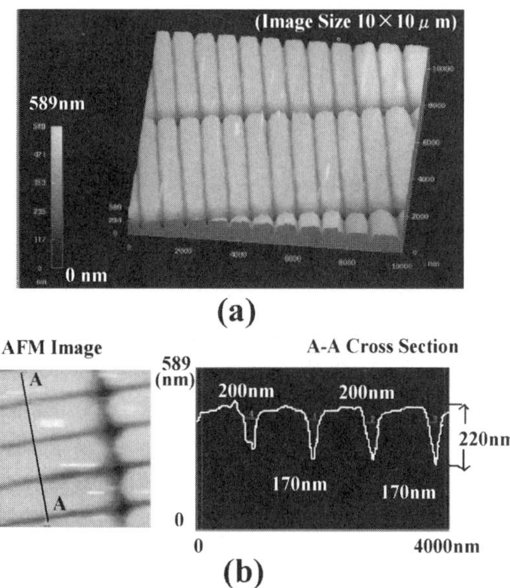

Figure 9.3 AFM image of (a) positive resist pattern obtained with near-field lithography; and (b) its profile.

with AFM measurement is in good agreement with that obtained by SEM measurement. However, the depth measured by AFM shows poor agreement with that of the cross-section measured by SEM because the images taken by the AFM frequently depend on local tip shape.

The near field is distributed at an extremely close range near the aperture and the slit, and decreases as $1/z^3$ near the aperture with increasing distance from the aperture,[13,14] where z is distance from the aperture. The distributed length can be regarded as the aperture-sized order. If the pattern size is reduced, the small gap between the mask and the substrate, ideally no gap, would be needed to transfer the pattern. To reduce far-field effect the use of hard contacts is effective. For this purpose, it seems that a mechanically deformable mask and substrate are suitable. One method is to use a deformable mask, such as a diaphragm[15] or plastic replica mold.[4,5] Another method is to use it as the substrate.[3] To obtain the hard contact an electrostatic force or a pressure is used. An attempt was made to transfer the subwavelength pattern on a resist-coated diaphragm that was fabricated by an anisotropic etching of silicon. The diaphragm had a thickness of 1 μm and a size of 1×1 mm². The apparatus used to obtain the small gap contact is shown in Figure 9.5. By evacuating the gap between the substrate and the mask, the diaphragm is pressed to contact the mask. Figure 9.6a shows the image of the silicon diaphragm on which the grid pattern is formed. When the diaphragm contacts the mask by evacuating the gap, a fine pattern is made, as shown in Figure 9.6b. However, a fine pattern cannot be obtained when contact is insufficient (Figure 9.6c).

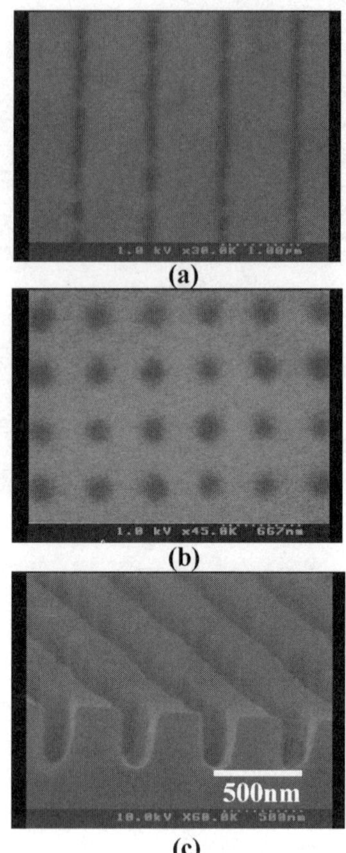

(a)

(b)

(c)

Figure 9.4 Example of (a) transferred grid; (b) dot patterns obtained by near-field lithography, and (c) cross section of the grid pattern on the resist.

9.4   Grid pattern transfer using polarized light

It is seen that when the far-field effect is decreased and localization of the near-field is enhanced on metal slits, the transferred pattern will become smaller. The near-field generated on metal slits is closely related to the polarization of incident lights as expected. Pattern transfer is demonstrated using near-field light that was created in S- and P-polarized light. After development, resist (positive type) patterns were observed by AFM. Figure 9.7 shows the transferred patterns using near-field light with different polarization. When the electric field of UV light is parallel to the metal slits (P-polarized light) (Figure 9.7a), the minimum size of the transferred patterns becomes small (minimum size is less than 150 nm). However, the transferred pattern is not clearly observed when the polarization direction is perpendicular to the metal slits

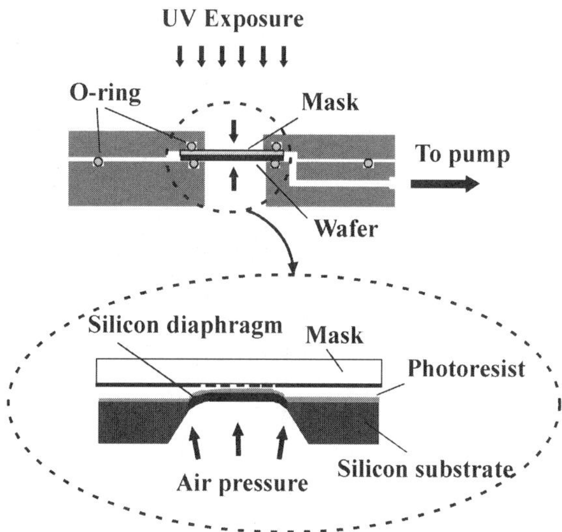

Figure 9.5 Schematic diagram of pattern transfer method by close contact of the silicon diaphragm with the mask.

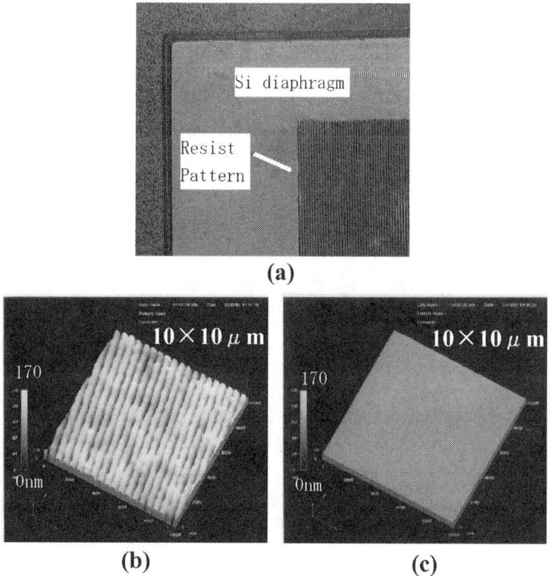

Figure 9.6 Resist pattern formed on the silicon diaphragm (a) when the diaphragm contacts with the mask by evacuating the gap a fine pattern is made; (b) the period of grids is 500 nm; and (c) no fine pattern could be obtained when the contact is poor.

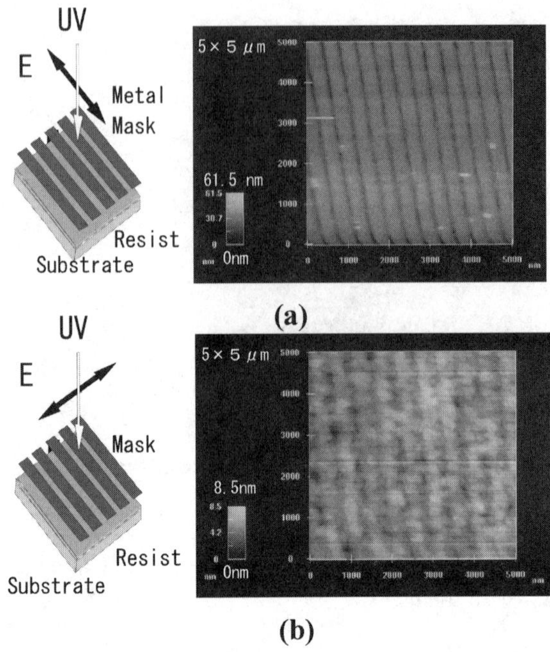

Figure 9.7 Resist pattern using polarized light: (a) UV light is parallel to the metal slits (P-polarized light); (b) the polarized direction is perpendicular to the metal slits (S-polarized light).

(S-polarized light) (Figure 9.7b). From simulation, it is shown that large-field enhancement effects appeared at the edges of the grating conductors for the P-polarized illumination.[5]

9.5 Conclusion

Subwavelength pattern transfer is demonstrated with optical near-field lithography. The minimum pattern size is beyond the diffraction limit of light, which suggests that the pattern is formed by near-field lithography. A mask having fine metal slits and dots with minimum size below one wavelength is made by EB and AFM-based lithography. The subwavelength pattern is transferred onto a positive resist using UV light (436 nm) by contact between the mask and the substrate. The near-field intensity decreases rapidly with increasing distance from aperture, so the gap between the mask and substrate must be small during exposure. A deformable mask or substrate is suitable to achieve close contact. It is found that polarized light decreases the minimum size of the transferred grid pattern. The limitation of this technique remains unclear at present. It is necessary to evaluate the near-field scattering and proximity effects for further investigations.

References

1. Wegscheider, S. et al., Scanning near-field optical lithography, *Thin Solid Films*, 264, 264, 1995.
2. Ghislain, L.P. et al., Near-field photolithography with a solid immersion lens, *Appl. Phys. Lett.*, 74, 501, 1999.
3. Ono, T. and Esashi, M., Subwavelength pattern transfer by near-field photolithography, *Jpn. J. Appl. Phys.*, 37, 6745, 1998.
4. Tanaka, S. et al., Printing sub-100 nanometer features near-field photolithography, *Jpn. J. Appl. Phys.*, 37, 6739, 1998.
5. Tanaka, S. et al., Simulation of near-field photolithography using the finite-difference time-domain method, *J. Appl. Phys.*, 89, 3547, 2001.
6. Rogers, J.A. et al., Generating ~90 nanometer features using near-field contact mode photolithography with an elastomeric phase mask, *J. Vac. Sci. Technol.*, B16, 59, 1998.
7. Betzig, E. et al., Near-field magneto-optics and high density data storage, *Appl. Phys. Lett.*, 61, 142, 1992.
8. Hosaka, S. et al., SPM-based data storage or ultrahigh density recording, *Nanotech.*, 8, 58, 1997.
9. Berger, S.D. et al., Projection electron beam lithography: a new approach, *J. Vac. Sci. Technol.*, B9, 2996, 1991.
10. Liddle, J.A. et al., Mask fabrication for projection electron-beam lithography incorporating the SCALPEL technique, *J. Vac. Sci. Tech.*, B9, 3000, 1991.
11. Chou, S.Y., Krauss, P.R., and Rentstrom, P.J., Imprint of sub-25 nm vias and trenches in polymers, *Appl. Phys. Lett.*, 67, 3114, 1995.
12. See for example, Rai-Choudhury, P., *Handbook of Microlithography, Micromachining, and Microfabrication, Vol. 1: Microlithography*, SPIE Press, Bellingham, WA, 1997.
13. Leviatan, Y., Study of near-zone fields of a small aperture, *J. Appl. Phys.*, 60, 1577, 1986.
14. English, R.E. and George, N., Diffraction from a small square aperture: approximate aperture fields, *J. Opt. Soc. Am.*, 5, 192, 1988.
15. Alkaisi, M.M. et al., Sub-diffraction-limited patterning using evanescent near-field optical lithography, *Appl. Phys. Lett.*, 75, 3560, 1999.

chapter ten

Optical near-field recording with a fabricated aperture array

One of the most promising applications of near-field optics is next-generation optical data storage. Optical memory with a high density and a high data transfer rate is in demand and needs an array of high throughput nanoscale light sources for writing and reading small bits on a medium. A systematic investigation of fabrication technology and optical performance of the fabricated apertures presented in previous chapters indicates that the structures could be applied for data storage. In this chapter, a hybrid structure of vertical cavity surface emitting lasers (VCSEL) and aperture array for optical near-field memory head, namely VCSEL/NSOM, is proposed. The first result of the fabrication and testing of writing and reading bits on a phase change medium (GeSbTe) using the fabricated structure is demonstrated.

10.1 Introduction

Together with magnetic memory, optical data storage plays a very important role in information technology. The density of magnetic memory has drastically increased about 60–100% annually and is now approaching its limitation due to the super paramagnetic effect that leads to a thermal instability of the recording bits at room temperature. As already mentioned, the density of optical memory is also limited by the diffraction effect. Size of recording optical bits (d) is dependent on spot size of the laser beam that depends on the laser wavelength (λ) and numerical aperture (NA) of the optical system, $d = 1.22\lambda/NA$ (NA = n sin (θ), n is the refractive index). To increase the density of optical data storage or reduce the size of recording bit, an obvious solution is to reduce the optical wavelength or increase the numerical aperture NA. However, optical storage with density of over 100 Gbit/inch^2 or with bit size smaller than 100 nm seems difficult to achieve.

To adapt to the increasing demand for data storage capacity in the twenty-first century, there is an urgent need to study and develop next-generation data storage. An effective next-generation data storage system must meet the following requirements:

- Write and read at a high speed (high data transfer rate)
- Create small and high packing density bits (high density)
- Read marks with the highest contrast (high ratio of signal/noise)
- Marks must be stable over time (long lifetime)
- Erase and rewrite capability
- Low cost

To meet these requirements, next-generation data storage needs contributions from many fields of science and technology. Novel recording techniques have been proposed by many groups in universities and companies. Details of these novel techniques can be found in the literature, for example, quantized magnetic recording[1]; magneto-optical recording using a solid immersion lens (SIL) and flying head technique[2,3]; near-field recording on magneto-optics (MO),[4] phase change (PC)[5] or photochromic materials[6]; using near-field light emitted from an aperture optical fiber probe; near-field recording using super-resolution near-field system (super-RENS)[7]; three-dimensional recording in photorefractive materials[8]; atomic force microscopy (AFM) based recording[9,10]; micro- or nanoheater integrated AFM probe array.[11-14]

The first demonstration of near-field recording is done by Betzig et al.[4] on a Pt/Co magneto-optical (MO) multilayer.[15] The write beam heats a small region under the aperture of the optical fiber NSOM tip, near the Curie temperature of MO film (300°C) resulting in the formation of a domain with a magnetization in the direction opposite to that of the surrounding materials without magnetic field. The recording pattern is then read back by the same probe with a lower power polarized laser beam. The Faraday or Kerr effects cause this polarization to be rotated slightly upon passage through or on reflection from the MO medium, respectively. The light is then collected with a conventional optical microscope and the rotation is measured with a polarizer/photomultiplier combination. The distance between the fiber tip and the medium is kept constant by the shear-force detection technique. A simplified schematic diagram of the recording system using near-field light on the MO media is shown in Figure 10.1. A density of 45 Gbits/in^2 was reported by Betzig.[4] This pioneering technique has opened up an area of research in optical data storage with near-field light. Some considerable disadvantages of the technique are:

- Limitation of the recorded magnetic domain due to the magnetic domain wall size
- Efficiency of the detecting signal using the incident light is low, so it is difficult to detect the domain at high speed with the reflected light
- Single probe was used so the data transfer rate is limited

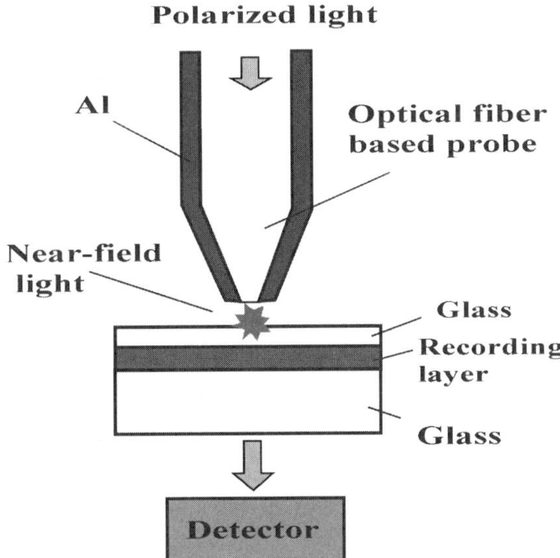

Figure 10.1 Schematic diagram of the near-field recording on the magneto-optic (MO) material using the near-field light emitted from an optical fiber-based NSOM tip (pioneered by Betzig et al.).

Hosaka et al.[5] have used near-field light emitted from the optical fiber NSOM probe for the formation and observation of 60 nm diameter phase change (PC) domain of a thin GeSbTe film. They achieved an ultra-high recording density of 170 Gbits/in^2. The efficiency of the detected light for reading can be improved compared to near-field recording on the MO medium because the reflectivity difference between the amorphous and crystalline phases in GeSbTe is around 30%. However, the data transfer rate is also limited due to the low optical throughput of the optical fiber and the single probe configuration as well as the shear-force technique for controlling the gap. A simplified schematic diagram of the recording system on the PC medium is shown in Figure 10.2.

To improve the data transfer rate in near-field recording with a single optical fiber NSOM probe, Hosaka et al. have also proposed a concept of near-field optical recording using a micromachined aperture on a flying head (see Figure 10.3).[15] The probe slider aperture was fabricated by photolithography and the focused ion beam cutting technique. A near-field reading of 1.56 Mbps was reported.

Terris et al.[2,3] proposed a novel recording technique using the solid immersion lens (SIL) and flying technique (see Figure 10.4). Using a SIL with refractive index n (n > 1), the spot size of a laser beam of d = 1.22/n.NA can be achieved. This means that when using the SIL technique bit size will be n times reduced compared to conventional optical data storage. With the SIL technique a density several times higher than conventional optical recording

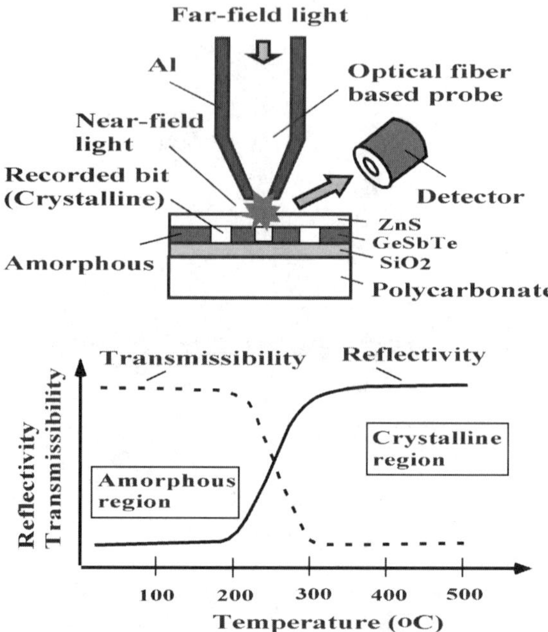

Figure 10.2 Schematic diagram of the near-field recording on the phase change (PC) material using the near-field light emitted from an optical fiber-based NSOM tip (pioneered by Hosaka et al.).

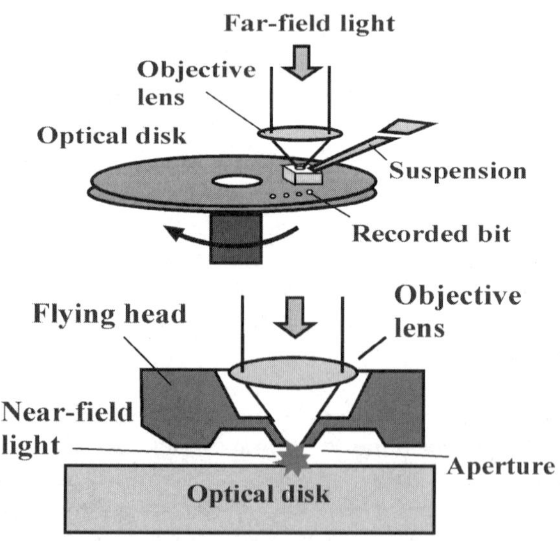

Figure 10.3 Schematic diagram of the near-field recording with single aperture on a flying head (pioneered by Hosaka et al.).

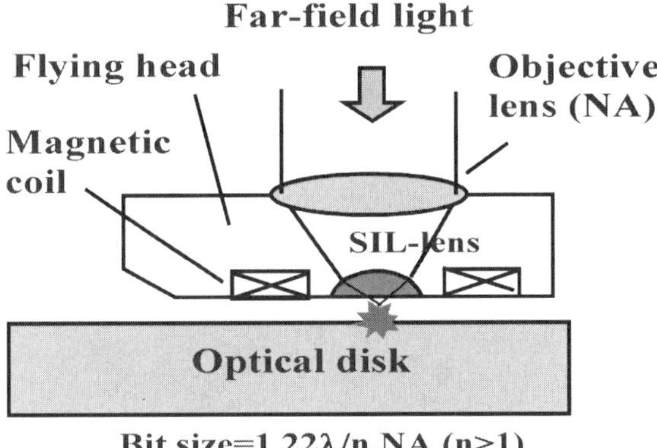

Figure 10.4 Schematic diagram of the near-field recording with solid immersion lens (SIL) lens (pioneered by Terris et al.).

and data transfer rate, one much higher than optical fiber based NSOM recording, can be expected.

Another novel optical near-field recording technique, super-resolution near-field structure (Super-RENS), has been intensively developed with the pioneering work of Tominaga et al.[7] In super-RENS, a near-field probe-like small aperture or scattering center is generated on a mask layer close to the recording layer on the disk by a focused laser beam. The transmitted light through the aperture or scattered light interacts with the recording layer placed within a near-field distance. Two kinds of super-RENS were developed: transparent aperture mode super-RENS using 15 nm thick Sb as an aperture layer; and light-scattering mode super-RENS using an AgO_x layer instead of Sb thin film. With these novel techniques, 100 nm bit size was successfully recorded and read out with 20–26 dB CNR (carrier noise ratio). The great advantage of this technique is that the near-field element is directly located on the recording disk. Therefore, a conventional optical recording system can be utilized.

In order to increase both packing density and the data transfer rate in optical near-field recording, other promising techniques of using near-field light at the aperture array for writing and reading bits were proposed by Goto et al.[16] and Ohtsu et al.[17] In this work, a concept of optical near-field recording head VCSEL/NSOM with a hybrid vertical cavity surface emitting laser array (VCSEL) and an aperture array at the apexes of SiO_2 tips on an Si diaphragm are also proposed. The concepts are a combination of the suggestions presented by Goto et al. and Ohtsu et al.[16,17] plus the advantages of this fabrication technique.

10.2 Concept of VCSEL/NSOM fabrication process

The concept of the hybrid VCSEL/NSOM memory head is shown in Figure 10.5.[18-19] An aperture array with a size of sub-50 nm is formed at the apexes

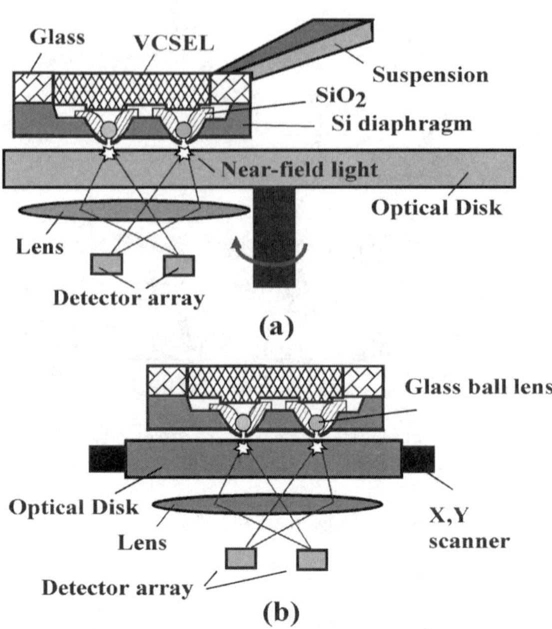

Figure 10.5 Concept of the hybrid VCSEL/NSOM near-field optical memory head; (a) flying head-type; (b) X, Y moving-type.

of the SiO_2 tips on an Si diaphragm that is developed from the Si process presented in Chapter 4. The Si diaphragm is anodically bonded with a glass base for handling. A VCSEL laser array and the aperture array will be aligned together with an adhesive agent. Glass ball lenses are put into the SiO_2 tips in between apertures and VCSEL lasers to enhance the near-field intensity at the aperture.[17] Far-field light from VCSEL laser array are focused onto the aperture array to form nanoscaled light sources for writing bits, for example, on phase change media. For reading bits, an array of photodetectors can be used as receivers and be aligned with the aperture array. The distance between the aperture and medium should be kept constantly closed by an air gap in flying head-type (Figure 10.5a) or a thin lubricating oil in X, Y moving-type (Figure 10.5b). Tracking may be utilized using one aperture in the array. With this proposal writing or reading multibits at one time can be utilized. The data transfer rate can be increased or the rotation speed of the disk can be reduced. Bit size can be as small as 20 nm, which is identical to the aperture diameter. Track pit can be also as small as 20–50 nm by rotating the structure a small angle to the tangential of the rotation of the disk. This structure is expected to serve as an ultra-high density optical disk system of terabits/inch^2. For writing a small bit, a small aperture with high optical throughput is very important. This technique is capable of fabricating an array of an ultra-small aperture with high optical throughput. This means that the structure is suitable for near-field data storage.

Analogy to the conventional optical recording, near-field recording can be used to write and read data on the magneto-optic (MO) or phase change (PC) medium. The PC medium was chosen in the first near-field recording testing because of the simplicity of the setup and the high efficiency of detectable light power in reading with PC (GeSbTe) medium.

The fabrication process is sketched in Figure 10.6. It is based on the low temperature oxidation and selective etching (LOSE) process presented in Chapter 4. A detailed explanation of the technique of forming an aperture as small as 20 nm at the apex of the SiO_2 tip can be found elsewhere.[20] First, pyramidal etch pits are defined on Si (100) wafer by oxidation, lithography, SiO_2 patterning and Si etching with a tetramethyl ammonium hydroxide TMAH (step a). Next, the wafer is thermally oxidized in wet oxygen at 950°C, to a thickness of about 1 μm (step b). Using a thin Cr film of about 100 nm as a mask, the top SiO_2 is partly etched as shown in step c. The wafer is next etched in the TMAH until forming the Si diaphragm with SiO_2 tips (step d). The wafer is subsequently dipped into a buffered-HF (BHF) for partly etching the SiO_2 until Cr protrusions are formed at the apexes of the tip array. After etching out of the Cr film, the aperture array is formed (step e). Next, a thin Al film (approximately 100 nm) is entirely deposited onto the back side of the Si diaphragm to form an opaque layer (step f). To concentrate more light into the aperture, an array of 3–10 μm diameter glass ball lenses is inserted on the upper side of the tips by spin coating (step f). For easy handling the wafer is anodically bonded with a Pyrex glass and the desired structure is released by dicing as shown in step h. Finally, the VCSEL laser array is glued on the Si diaphragm as shown in step i. VCSEL laser array is not yet so compact and expensive, but, it is expected that a compact and high yield VCSEL can be produced soon.

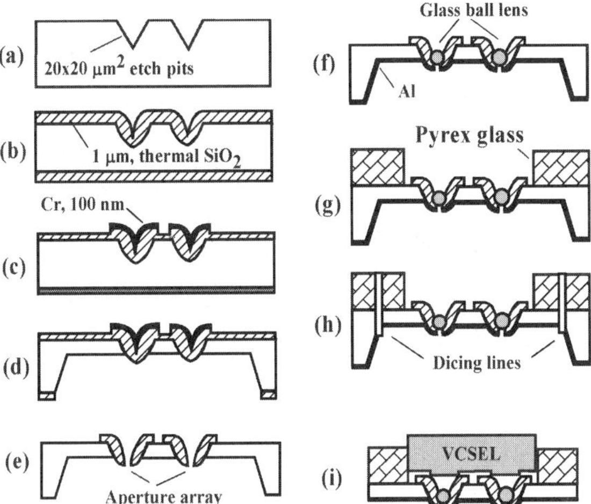

Figure 10.6 The fabrication process developed from the low temperature oxidation and selective etching technique.

10.3 *Results of fabrication and first result of recording*

SEM images of the fabricated aperture array on the Si diaphragm and close-up view of about 50 nm aperture is shown in Figures 10.7a and 10.7b, respectively. Typically, the aperture size in the range of 10–100 nm is obtained in the array. A glass ball lens is successfully inserted on the upper side of the tip as shown in Figure 10.8. With suitable condition of spin coating, the glass ball can be placed over 70% the number of tip. The fabricated aperture array was inspected using an inverted microscope. The light emitted through the aperture array with approximately 400 nm aperture was observed as shown in Figure 10.9.

For the purpose of recording, optical throughput (the ratio of optical powers of the near-field at the aperture and corresponding far-field light), is very critical. Since the opening angle of the SiO_2 tip in the array is large,

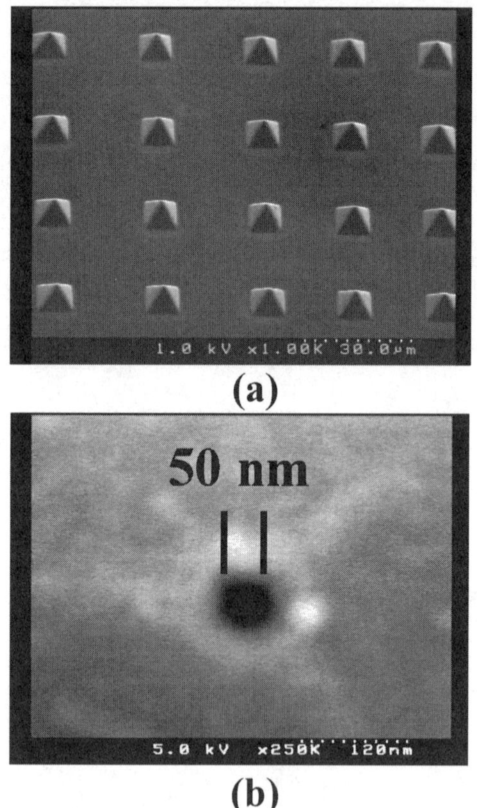

Figure 10.7 (a) SEM image of the microfabricated aperture array on Si diaphragm; (b) close-up view of 50-nm diameter aperture.

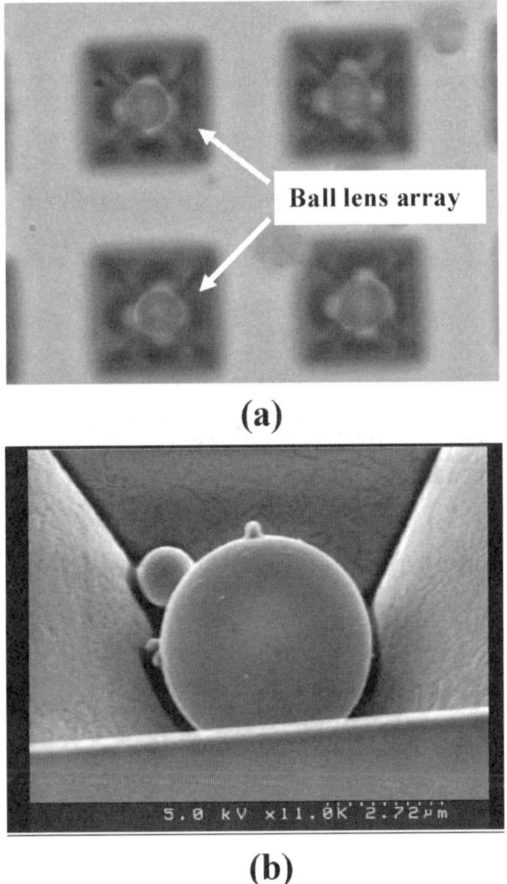

Figure 10.8 (a) Optical; and (b) SEM images of glass ball lenses on the upper side of the tip array that were produced by spin coating.

the optical throughput is very high. For 100 nm diameter aperture, approximately 1% optical throughput was confirmed by measurement and simulation as presented in Section 5.1.[21,22] With such high optical throughput, the proposed aperture array is suitable for data storage.

For the first testing of recording and reading ability of the proposed structure, one aperture of 400 nm in diameter in the array was used for writing and reading bits on PC medium using an AFM system. The recording testing system is schematically shown in Figure 10.10. A He-Cd laser of 442 nm wavelength and 10 mW power was focused onto the aperture with a 10 µm diameter ball lens located on it. Phase change medium with a 15 nm GeSbTe recording layer on glass substrate was located on a prism and an XYZ scanner.

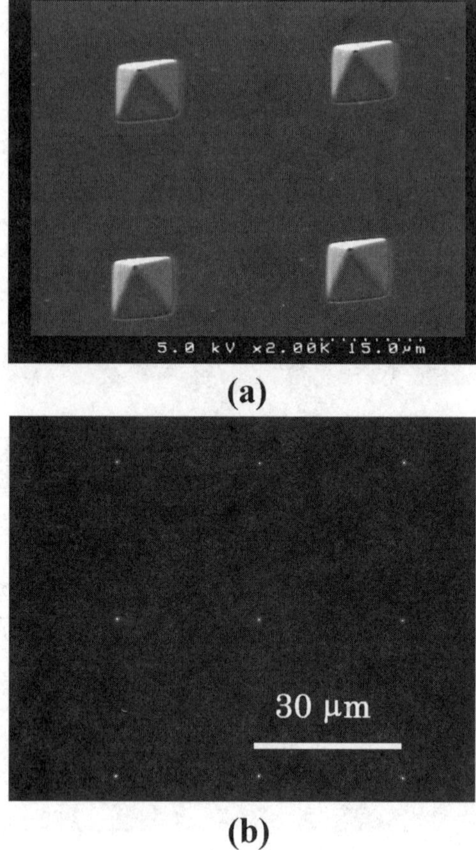

Figure 10.9 (a) SEM image of the fabricated aperture array with approximately 400-nm diameter; (b) optical image of the light coming out from the aperture array.

Near-field light at the aperture after interacting and transmitting through the medium is collected by a lens and detected by a photomultiplier tube in transmission. A photoimaging of the fabricated aperture array used for the recording testing is shown in Figure 10.11. Using the described setup, near-field recording on the PC medium was successfully demonstrated as shown in Figure 10.12. The cross-section is shown on the right side of Figure 10.12, which indicates that approximately 200–250 nm bit size was successfully recorded with a 400 nm aperture. The recorded bits appear as dark dots in the figure. This can be due to the refractive index and limited transmittivity of the crystalline of the recorded bits. Due to the limitation of the optical system and the scanning speed of the piezoelectric tube in the setup, the speed of reading and writing bits was not determined. However, since the optical throughput of the microfabricated aperture is very high, and multibits can write and read at the same time, the data transfer rate is expected to be very high. It was

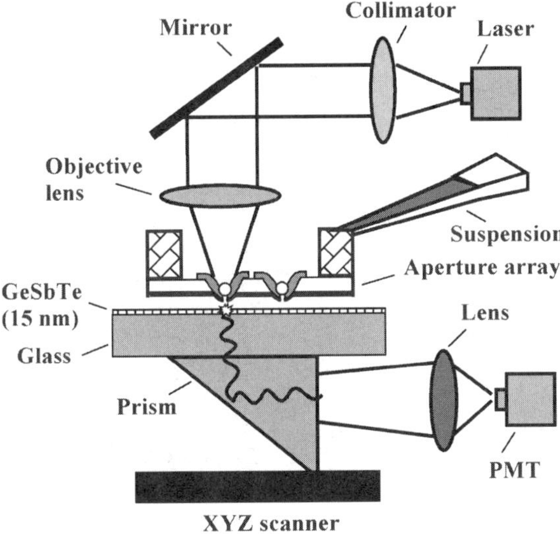

Figure 10.10 Schematic of the recording test using the fabricated aperture array based on the AFM system.

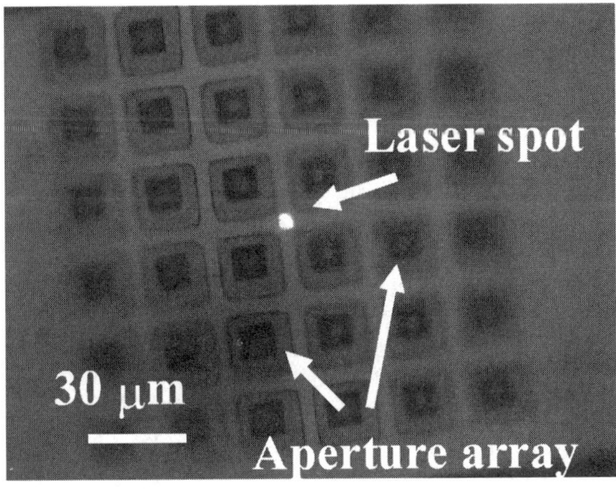

Figure 10.11 Optical image of the aperture array for recording testing.

expected that about 20 nm sized bits would be recorded on the PC medium using the sub-50 nm aperture corresponding to a density of terabits/inch^2. Future work should deal with recording, reading and erasing smaller bits on PC and MO media with a VCSEL laser. The structure may be capable of near-field recording with a density up to hundreds of bits per inch^2 or even terabit/inch^2 and data transfer up to Gbits/sec. The near-field light can also be utilized for photon mode recording on photochromic materials.

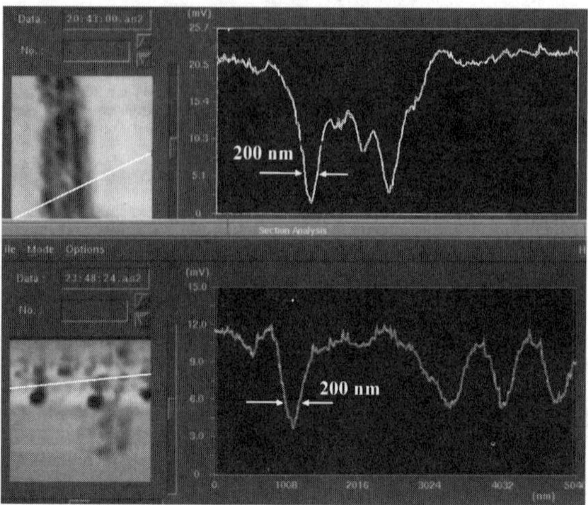

Figure 10.12 Using an aperture of 400 nm a 200–250-nm bit size on 20 nm GeSbTe PC medium was recorded (black dots) in the aperture array using the setup shown in Figure 10.10, reading bits were recorded with a lower laser light.

10.4 Discussion

Increasingly efficient data storage is a huge market and requires a correspondingly enormous investment. Micro- and nanofabrication plays a very important role in the fabrication of writing, reading and erasing heads. To achieve high speed, high-density storage, a novel medium also plays an important role.

Magnetic recording still plays a leading role because of high speed; however, it seems difficult to achieve density as high as terabits/inch². Among other novel recording techniques, near-field recording using an array of high optical throughput and a super-small aperture array is very promising. A hybrid structure of optical fiber bundle and aperture array is also one of the good choices. A probe array using an AFM tip or integrated nanoheater is also expected to be realized soon.

References

1. Chou, S.Y., Krauss, P.R., and Kong, L., Nanolithographically defined magnetic structures and quantum magnetic disk, *J. Appl. Phys.*, 79, 6101, 1996.
2. Terris, B.D. et al., Near-field optical data storage using a solid immersion lens, *Appl. Phys. Lett.*, 65, 388, 1994.
3. Terris, B.D., Mamin, H.J., and Rugar, D., Near-field optical data storage, *Appl. Phys. Lett.*, 68, 141, 1996.
4. Betzig, E. et al., Near-field magneto-optics and high density data storage, *Appl. Phys. Lett.*, 61, 142, 1992.
5. Hosaka, S. et al., Phase change recording using a scanning near-field optical microscope, *J. Appl. Phys.*, 79, 8082, 1996.

6. Jiang, S. et al., High localized photochemical processes in LB films of photo-chromic material by using a photon scanning tunneling microscope, *Opt. Commun.*, 106, 173, 1994.

7. Tominaga, J., Nakano, T., and Atoda, N., An approach for recording and readout beyond the diffraction limit with an Sb thin film, *Appl. Phys. Lett.*, 73, 2078, 1998.

8. Ueki, H., Kawata, Y., and Kawata, S., Three dimensional optical bit-memory recording with a photorefractive crystal: analysis and experimental, *Appl. Opt.*, 35, 2457, 1996.

9. Kado, H. and Tohda, T., Nanometer-scale recording on chalcogenide films with an atomic force microscope, *Appl. Phys. Lett.*, 66, 2961, 1995.

10. Cooper, E.B. et al., Terabit-per-square-inch data storage with the atomic force microscope, *Appl. Phys. Lett.*, 75, 3566, 1999.

11. Binnig, G. et al., Ultrahigh-density atomic force microscopy data storage with erase capability, *Appl. Phys. Lett.*, 74, 1329, 1999.

12. Takimura, N. et al., Heater integrated micro probe for high density data storage, *Tech. Dig. of the 17th Sensor Symposium*, Japan, 423, 2000.

13. Lee, D.W. et al., Fabrication of microprobe array with sub-100 nm nano-heater for nanometric thermal imaging and data storage, *Tech. Dig. of 14th IEEE International MEMS-01*, Interlaken, Switzerland, 204, 2001.

14. Ono, T. et al., Micromachined probe for high density data storage, *Tech. Dig. of the 4th Pacific Rim Conference on Lasers and Electro-Optics*, Makuhari Messe, Chiba, Japan, II-542, 2001.

15. Issiki, F., Itoh, K., and Hosaka, S., 1.5-Mbit/s direct readout of line-and-space patterns using a scanning near-field optical microscopy probe slider with air-bearing control, *Appl. Phys. Lett.*, 76, 804, 2000.

16. Goto, K., Proposal of ultrahigh density optical disk system using a vertical cavity surface emitting laser array, *Jpn. J. Appl. Phys.*, 37, 2274, 1998.

17. Lee, M.B. et al., Silicon planar-apertured probe array for high density near-field optical storage, *Appl. Opt.*, 38, 3566, 1999.

18. Minh, P.N. et al., High throughput optical near-field aperture array for data storage, *Tech. Dig. of the 14th IEEE International Conference MEMS' 01*, Interlaken, Switzerland, 309, 2001.

19. Minh, P.N. et al., Near-field recording with high optical throughput aperture array, *Sens. Actu. A*, forthcoming.

20. Minh, P.N., Ono, T., and Esashi, M., Nonuniform silicon oxidation and application to the fabrication of aperture for near-field scanning optical microscopy, *Appl. Phys. Lett.*, 75, 4076, 1999.

21. Minh, P.N., Ono, T., and Esashi, M., High throughput aperture near-field scanning optical microscopy, *Rev. Scientific Instru.*, 71, 3111, 2000.

22. Minh, P.N. et al., Near-field optical apertured tip and modified structures for local field enhancement, *Appl. Opt.*, 40, 2479, 2001.

part 3

Conclusion

chapter eleven

Future aspects and conclusions

11.1 Outlook for the future

At the beginning, near-field optics was devoted for ultra-high resolution optical microscopy. However, it has been quickly realized that near-field optics can open a new frontier in nanooptical science and technology. With near-field optics, physicists, chemists, biologists, and engineers have a tool in hand to observe, understand, and control nanometer scale structures. Applications using near-field light are increasingly expanding. Some of those applications are as follows:

- Ultrahigh-resolution optical imaging spectroscopy
- Luminescent spectrum of single quantum wells, quantum wires and quantum dot in semiconductor
- Fluorescent spectrum of biological cells, single molecules or DNA
- Surface plasmon and local field enhancement
- Near-field femto second study
- Atom trapping and manipulation
- High-resolution magnetic imaging using the magneto-optic Kerr or Faraday effects
- Nano-Raman scattering or infrared spectroscopy for chemical identity, material phase or stress
- Subwavelength photolithography
- Nanofabrication and deposition using near-field light
- Ultra-high density near-field optical data storage

Theoretical works on near-field optics and interaction of near-field light with materials on a nanometric scale are also becoming more applicable.

To effectively utilize the near-field light, it is very important to fabricate a probe with high reproducibility, both of the shape and size of the tip and the aperture with as high optical throughput as possible. Probe array is essential for many applications. The current optical fiber technology is advanced in some cases, however, many drawbacks still exist. Si micromachining has been shown to be a powerful tool in the fabrication of near-field optical elements. It is confidently expected that microfabricated NSOM probes will overcome most of the drawbacks of current optical fiber based probes.

11.2 Conclusions

This work presents several novel technological solutions for fabrication of ultra-small apertures and aperture array for near-field applications, investigation of these optical properties and demonstration of several actual applications in near-field optical imaging photolithography and data storage. This work can be summarized as follows:

- A simple and effective technique for manufacturing apertures with diameters as small as 20 nm at the apex of the SiO_2 tip or tip array for optical near-field applications has been found. The technique is based on the fundamental process of Si micromachining technology in a batch production and called low temperature oxidation and selective etching (LOSE). The reproducibility of the technique is very high due to the advantages of Si micromachining. This is a key point of this work. It is expected that several new devices for application in nanotechnology can be developed with the technique.
- The fabricated probes show very high optical throughput due to the geometric structure of the SiO_2 tip that was grown at a low temperature (around 1% throughput was confirmed by measurement and simulation for 100 nm aperture). This optical throughput is significantly high compared to the conventional fiber tip (10^{-5}–10^{-6} for 100 nm aperture). This is one of the most important features of the structure for optical near-field applications.
- The near-field light is strongly located at the fabricated aperture as a bright nanolight source with confinement and highly controllable polarization.
- Simulation with FDTD technique was in agreement with the measurement. Since the geometrical structure (shape and size) of the tip and the aperture is well defined, the simulation is easily adaptable.
- The fabricated probes are based on the Si cantilevers that can be operated in AFM/NSOM or capacitive–AFM/NSOM. The probes are very reliable because of the low stiffness of the Si cantilever in a vertical direction. Since the tip–sample distance is kept constant by the AFM feedback technique, the feedback is much faster and more reliable than the shear-force feedback technique in optical fiber based NSOM probe. The conventional AFM system can be used for this probe. With the presented aperture probes and instrumentations, it is expected to be utilized in bio-imaging and spectroscopy as well as subwavelength photolithography.

- Optical near-field imaging of several surfaces with resolution of 15 nm ($\lambda/52$) was observed with the fabricated probe.
- Optical near-field lithography with narrow apertures or slits was demonstrated using a hard contact lithography technique.
- Near-field recording on phase change medium was demonstrated with fabricated aperture array. It is expected to achieve tera bits/inch^2 by utilizing aperture array with size of aperture around 20–30 nm. Due to structure array it is feasible to increase the data transfer rate up to an order of Gbps by utilizing multirecording and reading.
- Novel probes of coaxial NSOM with a metal nanowire, single carbon nanotube (CNT) or Ag embedded particle at the center of the aperture for enhanced optical throughput were developed.
- A hybrid structure of optical fiber and an apertured cantilever, and optical fiber bundle and aperture array for high speed storage or near-field processing has been developed. With this structure, advantages of both optical fiber and apertured cantilever can be utilized. Moreover, disadvantages of optical fiber and apertured cantilever can be eliminated.
- A simple and effective method of forming a microlens array at the fiber's core has also been developed. The process can be done in a mass production. We expected the technique could find applications in other fields using optical fiber based tools.

The technique of forming an ultra-small aperture is very important for many novel devices that work in the nanoworld, such as:

- Fabrication of an integrated nanothermo couple or nanoheater at the apex of the SiO_2 tip by filling metal through the nanohole for nanoscanning thermal microscopy for investigation of thermal properties of surface at nanometer scale.
- The probe with integrated nanoheater can make an array and an IC integrated circuit for creating pits on the medium for high density data storage.
- Integrated waveguide on the cantilever for a compact NSOM probe.
- Electron field emission devices with integrated electrostatic lens can be developed for electron multibeam lithography or other electron beam based application.
- Novel probes for analyzing material at nanoscale such as: nano-Raman or nano-Auger, can be developed.

This book is expected to give the reader useful information with several examples that have been demonstrated by this group. It is hoped that readers can develop other fabrication technique and applications based on the fabrication technology presented here.

Index

A

B

C